G Proteins and Signal Transduction

Society of General Physiologists Series • Volume 45

G Proteins and Signal Transduction

Society of General Physiologists • 43rd Annual Symposium

Edited by
Neil M. Nathanson
University of Washington
and
T. Kendall Harden
University of North Carolina
at Chapel Hill

Marine Biological Laboratory
Woods Hole, Massachusetts

6–9 September 1989

© The Rockefeller University Press
New York

Copyright © 1990 by The Rockefeller University Press
All rights reserved
Library of Congress Catalog Card Number 90-81669
ISBN 0-87470-046-9
Printed in the United States of America

Contents

Preface		vii
Chapter 1	Networking Ionic Channels by G Proteins *Arthur M. Brown, Atsuko Yatani, Antonius M. J. VanDongen, Glenn E. Kirsch, Juan Codina, and Lutz Birnbaumer*	1
Chapter 2	Modulation of Neuronal Calcium Currents by G Protein Activation *A. C. Dolphin and R. H. Scott*	11
Chapter 3	A Comparison of the Roles of Purified G Protein Subunits in the Activation of the Cardiac Muscarinic K^+ Channel *Chaya Nanavati, David E. Clapham, Hiroyuki Ito, and Yoshihisa Kurachi*	29
Chapter 4	Modulation of M Current in Frog Sympathetic Ganglion Cells *Martha M. Bosma, Laurent Bernheim, Mark D. Leibowitz, Paul J. Pfaffinger, and Bertil Hille*	43
Chapter 5	Regulation of Phospholipase C *Andrew J. Morris, Gary L. Waldo, Jose L. Boyer, John R. Hepler, C. Peter Downes, and T. Kendall Harden*	61
Chapter 6	Structure/Function Relationships of Ras and Guanosine Triphosphatase-activating Protein *Jackson B. Gibbs, Michael D. Schaber, Victor M. Garsky, Ursula S. Vogel, Edward M. Scolnick, Richard A. F. Dixon, and Mark S. Marshall*	77
Chapter 7	Regulation of G Protein-coupled Receptors by Agonist-dependent Phosphorylation *Jeffrey L. Benovic, James J. Onorato, Marc G. Caron, and Robert J. Lefkowitz*	87
Chapter 8	Molecular Genetics of Signal Transduction by Muscarinic Acetylcholine Receptors *Mark R. Brann, Jürgen Wess, and S. V. Penelope Jones*	105
Chapter 9	Chimeras of the G_s α Subunit That Had the NH_2- or COOH-Terminal Sequence Substituted for the Corresponding Region of the G_i α Subunit Constitutively Activate Adenylyl Cyclase *Lynn E. Heasley, N. Dhanasekaran, Sunil K. Gupta, Shoji Osawa, and Gary L. Johnson*	117

Chapter 10	G Protein-linked Signal Transduction in Aggregating *Dictyostelium* *Geoffrey S. Pitt, Robert E. Gunderson, Pamela Lilly, Maureen B. Pupillo, Roxanne A. Vaughan, and Peter N. Devreotes*	125
Chapter 11	Immunological and Molecular Biological Analysis of the Regulation and Function of Muscarinic Acetylcholine Receptors and G Proteins *Neil M. Nathanson*	133
Chapter 12	Structural and Functional Studies of the G_o Protein *Eva J. Neer*	143
Chapter 13	The Molecular Components of Olfaction *Randall R. Reed*	153
Chapter 14	*Drosophila* G Proteins *James B. Hurley, Stuart Yarfitz, Nicole Provost, and Sunita de Sousa*	157
Chapter 15	G Protein Coupling of Receptors to Ionic Channels and Other Effector Systems *Lutz Birnbaumer, Atsuko Yatani, Antonius M. J. VanDongen, Rolf Graf, Juan Codina, Koji Okabe, Rafael Mattera, and Arthur M. Brown*	169
Chapter 16	Antibodies against Synthetic Peptides as Probes of G Protein Structure and Function *Allen M. Spiegel, William F. Simonds, Teresa L. Z. Jones, Paul K. Goldsmith, and Cecilia G. Unson*	185
Chapter 17	Mechanisms of G Protein Action: Insight from Reconstitution *Paul C. Sternweis and Iok-Hou Pang*	197
List of Contributors		207
Subject Index		211

Preface

The GTP-binding regulatory proteins, or G proteins, are a family of signal transduction proteins that couple hormone and neurotransmitter receptors to effector proteins involved in the generation of intracellular and transmembrane signals. The G proteins were originally identified as transmembrane mediators of the activation of the light-sensitive cyclic GMP phosphodiesterase by rhodopsin and of adenylate cyclase by hormone receptors. This family of signaling proteins was generally thought to consist of a small number of heterotrimers with α subunits that were substrates for ADP-ribosylation by bacterial toxins such as cholera toxin or islet-activating protein (pertussis toxin). Both the number and functions of G proteins have grown dramatically in recent years. The most common mechanism for regulation of ion channel activity by neurotransmitter receptors is likely to involve G protein–mediated coupling, and receptor-mediated activation of phospholipase C also involves a G protein. In addition, G proteins have been implicated in such diverse functions as regulation of cell growth, exocytosis, fertilization, and oncogenesis.

This increased knowledge of the diversity of of G protein function has been matched with an increase in the number of known G proteins. The application of molecular cloning techniques has allowed the identification of multiple forms of both the heterotrimeric G proteins and the low molecular weight ras-like proteins; the elucidation of the specificity of the functional roles of these isoforms remains a problem for the future. Similarly, the increased diversity of both the receptors and effectors known to interact with the G proteins raises questions about the mechanisms regulating the expression as well as the functional integration of components in the G protein–mediated signal transduction pathways.

We are grateful to the Society of General Physiologists for devoting their 43rd symposium to the topic of G proteins and signal transduction. We would like to thank Dick Tsien for originally suggesting that we organize this symposium, Jane Leighton both for ensuring that we actually did everything that we needed to for the meeting to take place and for carrying out a myriad of tasks herself, and Susan Lupack for editorial assistance in the preparation of the abstracts and these symposium proceedings. Finally, we would like to acknowledge the following organizations for providing financial support for the symposium: Genentech, Inc., ICI Pharmaceuticals Group, National Institutes of Health, National Science Foundation, Office of Naval Research, US Army Medical Research and Development Command, and US Army Research Office.

Neil M. Nathanson
T. Kendall Harden

Chapter 1

Networking Ionic Channels by G Proteins

Arthur M. Brown, Atsuko Yatani, Antonius M. J. VanDongen,
Glenn E. Kirsch, Juan Codina, and Lutz Birnbaumer

*Department of Molecular Physiology and Biophysics and the
Department of Cell Biology, Baylor College of Medicine, Houston,
Texas 77030*

Introduction

Signal transducing, guanine nucleotide–binding (G) proteins are heterotrimers that couple receptors to effectors within the plasma membrane of eukaryotic cells (Birnbaumer et al., 1987; Gilman, 1987). As couplers, the G proteins are thought to amplify and terminate the effects of receptor occupancy by agonist. Also, one type of receptor may interact with more than one type of G protein, allowing for multiple effects (Ashkenazi et al., 1987), and different receptors may interact with one or more G proteins, allowing for summation of action on a single effector (Birnbaumer et al., 1987; Gilman, 1987; Fig. 1). The receptors seem to be members of a family having seven transmembrane segments (Dixon et al., 1986). The membrane effectors, such as adenylyl cyclase (AC), cyclic GMP phosphodiesterase, and phospholipase C, then

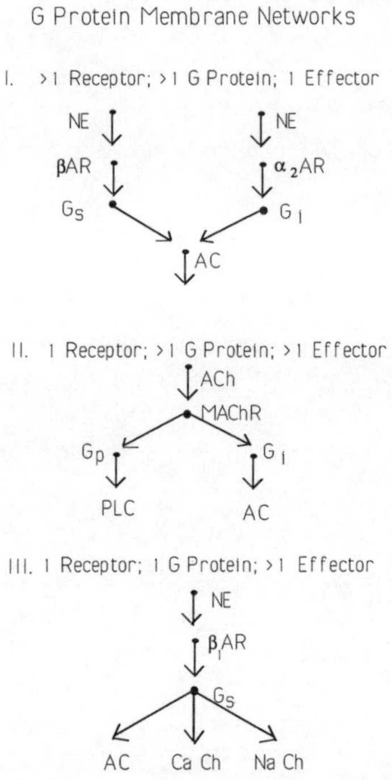

Figure 1. One or more G proteins may interact with one or more receptors and one or more effectors to produce complex membrane networks. Three examples are shown. *NE*, norepinephrine; *βAR*, beta-adrenoreceptor; $\alpha_2 AR$, alpha2-adrenoreceptor; G_s, stimulatory G protein regulator of AC; G_i, inhibitory G protein regulator of AC; *ACh*, acetylcholine; *MAChR*, muscarinic acetylcholine receptor; *PLC*, phospholipase C; *CaCh*, dihydropyridine-sensitive, high threshold Ca^{2+} channel; *NaCh*, tetrodotoxin-sensitive Na^+ channel.

change the levels of cytoplasmic second messengers such as cAMP, cCMP, or Ca^{2+}, which in turn produce a complex cellular response. The rates at which cellular responses occur are necessarily limited by the rates at which the changes in cytoplasmic messengers occur and for this reason need not be tightly tied to the presence of agonist. However, such an indirect sequence is not the only one by which G proteins and their effectors alter cellular function. Recently, ionic channels, which are known to be targets for cytoplasmic messengers and hence indirect targets of G proteins, have been shown to be direct or, more correctly, membrane-delimited targets for G proteins (Codina et al., 1987a, b; Yatani et al., 1987a, c; Brown and Birnbaumer, 1988). Because ionic channels are remarkable amplifiers of ionic fluxes (1,000–1,000,000 ions/s), global changes in cellular membrane potential occur quickly. Furthermore, a single G protein can have more than one effector as its target; in fact cardiac G_s has

AC, and Na$^+$ and Ca^{2+} channels as direct targets (Yatani et al., 1987c, 1988a; Imoto et al., 1988; Mattera et al., 1989a; Schubert et al., 1989a; Yatani and Brown, 1989). Therefore, G proteins act as switches that alter potential-dependent cellular functions such as contraction or secretion quickly via membrane pathways and more slowly via cytoplasmic pathways. In this paper we will examine the evidence for dual G protein pathways that regulate the function of Ca^{2+} and Na$^+$ channels.

Indirect G Protein Regulation of Ionic Channels

Examples of G protein-linked cytoplasmic pathways that regulate ionic channels are phosphorylation of Ca^{2+} and Na$^+$ channels by cAMP-dependent protein kinase in heart (Fig. 2), phosphorylation of Ca^{2+} channels by protein kinase C in heart and neurons (Dunlap et al., 1987; Lacerda et al., 1988), cGMP regulation of the nonselective cation channel in photoreceptors (Stryer, 1986), and Ca^{2+} regulation of K$^+$ channels, especially in smooth muscle. In one of the best studied examples, namely the β-adrenoreceptor-G_s-AC-cAMP-protein kinase A-cardiac Ca^{2+} channel pathway, each step in the pathway has been identified by blocking or bypassing preceding steps and by

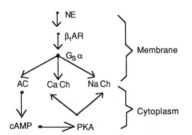

Figure 2. Membrane and cytoplasmic pathways by which G_s links β-adrenoreceptors to ionic channels. Abbreviations as in the legend of Fig. 1.

showing correlations between different steps (Kameyama et al., 1985; Trautwein and Cavalie, 1985). Thus cAMP and cAMP-dependent protein kinase applied intracellularly have the same effect as β-agonists applied extracellularly. The dose-response curve for the Ca^{2+} current-β-agonist (isoproterenol or ISO) response, for the Ca^{2+} current-cAMP response, and for the Ca^{2+} current-protein kinase A response were predicted from the changes in the cytoplasmic messenger levels produced by ISO. While the sequence is known, the rate at which Ca^{2+} currents are changed by these cytoplasmic messengers after β-agonist activation, appears to require several seconds (Yatani and Brown, 1989).

The question of basal phosphorylation or basal modulation by some other component has not been settled although the cAMP-dependent protein kinase does not seem to be importantly involved. The situation is complicated because this kinase has other substrates besides Ca^{2+} channels, including, in heart, Na$^+$, delayed rectifier K$^+$ (Giles and Imaizumi, 1988), and pacemaking (I_f) (Yanagihara and Irisawa, 1980; DiFrancesco and Tromba, 1988) channels, phosphorylase kinase, and phosphorylase A. Protein phosphatases are also important but their regulatory mechanisms have not been defined.

Direct G Protein Regulation of Ionic Channels

The membrane-delimited nature of direct pathways was demonstrated principally using the inside-out configuration of the patch clamp or through the incorporation of plasmalemmal membrane vesicles into planar lipid bilayers. In these circumstances the cytoplasm has been eliminated and the ionic channel responses were reconstituted through either receptor–G protein pathways or preactivated (with GTPγS usually) by specific G proteins. So far, a direct connection between a G protein and ionic channels, similar to that between G_s and AC or G_t and cGMP PDE, has not been shown and it is possible that some other membrane component is interposed between the G protein and the ionic channel.

The best studied pathway is that between the muscarinic M_2 atrial receptor, the G protein called G_k (Breitweiser and Szabo, 1985; Brown and Birnbaumer, 1988), and the specific atrial K^+ channel gated by this G protein. This is an example of obligatory G protein gating of an ionic channel. The α subunit mimicked the holoprotein effect and neither G_s nor G_t were effective. Other examples in this category are shown in Table I. Surprisingly we have found that all three $G_i\alpha$'s are equipotent, suggesting that specificity may reside in muscarinic receptor–G protein coupling (Yatani et al., 1988b). More recently it has been suggested that this channel may be modulated by arachidonic acid or its 5-hydroxy-eicosatetroic acid-lipoxygenase metabolites after dimeric $\beta\gamma$ activation of phospholipase A_2 although these effects are not involved in the muscarinic response (Kim et al., 1989; Kurachi et al., 1989).

The cardiac Ca^{2+} channel is another example of a membrane-delimited interaction between an ionic channel and specific G protein. In fact the same high threshold, dihydropyridine-sensitive Ca^{2+} channel that is phosphorylated by the G_s cytoplasmic pathway is directly activated by G_s (Yatani et al., 1987b, 1988a; Imoto et al., 1988; Mattera et al., 1989b; Yatani and Brown, 1989). Unlike obligatory G protein gating, G_s acts as a modulator in this case; only membrane depolarization actually opens these channels. When this effect was first observed, an immediate question was whether the same $G_s\alpha$ was an effector for both AC and Ca^{2+} channels. We tested this using recombinant $G_s\alpha$'s made in *Escherichia coli* and found that three of the four $G_s\alpha$'s were equally effective on AC and Ca^{2+} channels (Mattera et al., 1989a). Hence, a single $G_s\alpha$ may have more than one effector. This possibility had been raised earlier for AC and a Mg^{2+} transporter in liver cells (Maguire and Erdos, 1980). Next we tested whether such modulatory effect might be present for other channels. An interesting candidate was the voltage-dependent Na^+ current in cardiac cells which is responsible for propagating the cardiac impulse. There was circumstantial evidence that these channels, which are not usually considered to be highly regulated, might also be modulated by the β-adrenoreceptor-G_s pathway (Hisatome et al., 1985). We found that, just as for cardiac Ca^{2+} channels, G_s acted on cardiac Na^+ channels by dual direct and indirect pathways, although in this case both effects were inhibitory rather like the effects of local anesthetics. Thus it appears that in heart G_s regulates AC, and Ca^{2+} and Na^+ channels (Fig. 2) by dual pathways. Another channel that is a likely candidate for such dual regulation is the delayed rectified K^+ channel in heart (Walsh and Kass, 1988). The functional significance of this coordinate regulation is unknown at present, however, one example could be the case of myocardial ischemia with the accompanying increase in cardiac sympathetic tone. Depolarization because of K^+ accumulation would tend to bring β-adrenergic-G_s inhibition of cardiac Na^+ currents into play while

Ca^{2+} currents would be enhanced. Hence propagation of the cardiac impulse in this situation would be by the slow Ca^{2+} channels. The cytoplasmic messengers would tend to exaggerate these effects as well as bringing other effects into play.

We propose therefore that G_s organizes the β-adrenergic spatially by activating several effectors. We also propose that G_s organizes the β-adrenergic response temporally by its direct membrane messenger effects and by its indirect cytoplasmic

TABLE I
Ionic Channels Gated Directly by Proteins

Gα protein	Channel	Receptor	Tissue	References
α_i-3	K^+, 40 pS, IR	M_2, (ACh)	Atrium	Yatani et al., 1987a
				Codina et al., 1987b
				Mattera et al., 1989b
				Yatani et al., 1988
α_i-3	K^+, 55 pS, ?R	M_2, (ACh), SS	GH$_3$	Codina et al., 1987a
				Yatani et al., 1987c
α_o	K^+, 55 pS, NR	Uknown	Hippocampus	VanDongen et al., 1988
	K^+, 38 pS, NR	Uknown	Hippocampus	VanDongen et al., 1988
	K^+, 38 pS, IR	5-HT1A	Hippocampus	VanDongen et al., 1988
	K^+, 13 pS, NR	Unknown	Hippocampus	VanDongen et al., 1988
α_i-1	K^+, 40 pS, IR	M_2, (ACh)	Atrium	Yatani et al., 1988
α_i-2	K^+, 40 pS, IR	M_2, (ACh)	Atrium	Yatani et al., 1988
α_s	Ca^{2+}, DHP-sensitive, 25 pS	β-AR	Atrium, ventricle	Yatani et al., 1987b Imoto et al., 1988
Splice variants of α_s	Ca^{2+}, DHP-sensitive, 10 pS	β-AR	Skeletal muscle	Yatani et al., 1988
α_s	Na^+, TTX-sensitive	β-AR	Atrium, ventricle	Schubert et al., 1989b
α_i-3	Epithelial Na^+	Unknown	A6 cells	Light et al., 1989
α_i-3	Cation channel	Unknown	Kidney	Schwiebert et al., 1989
				Cantiello et al., 1989
α_i-3	K^+_{ATP}	Unknown	RIN	Eddlestone et al., 1989
α_o	K^+_{ATP}	Unknown	Skeletal muscle	Parent and Coronado, 1989
α_i-3	Cl^-	Unknown	Kidney	Schwiebert et al., 1989
				Cantiello et al., 1989
Unknown	K^+_{Ca}, 260 pS	β-AR	Myometrium	Ramos-Franco et al., 1989
Unknown	Ca^{2+}, T-type	Unknown	Dorsal root ganglion	Ma and Coronado, 1988

IR, inwardly rectifying K^+ channel; *NR*, normal outwardly rectifying K^+ channel; *SS*, somatostatin; *β-AR*, β-adrenoreceptor; *RIN*, rat insulinoma.

messenger effects. One possible manifestation of this would be a biphasic response of cardiac Ca^{2+} currents to β-agonist stimulation with one component occurring at the same rate as the muscarinic response and the other occurring at the slower rates expected for cytoplasmic messenger effects. We tested for this using a concentration jump method that allowed us to change test concentrations in ~10 ms and found a biphasic response that was similar to our expectations. We presume that the branch point is G_s because bypassing G_s using, for example, forskolin to directly stimulate AC or isobutylmethylxanthine (IBMX) to elevate cAMP levels, produced only the slow monophasic cytoplasmic response. When forskolin or IBMX produced maximal effects, an additional β-agonist effect could still be produced, presumably by the direct pathway (Fig. 3). Similar results have now been reported in other laboratories (Trautwein, W., A. Cavalie, T. J. A. Allen, Y. M. Schuba, S. Pelzer, and D. Pelzer, manuscript submitted for publication). A biphasic response due to a similar mechanism was also found for β-adrenergic modulation of cardiac Na^+ currents. One

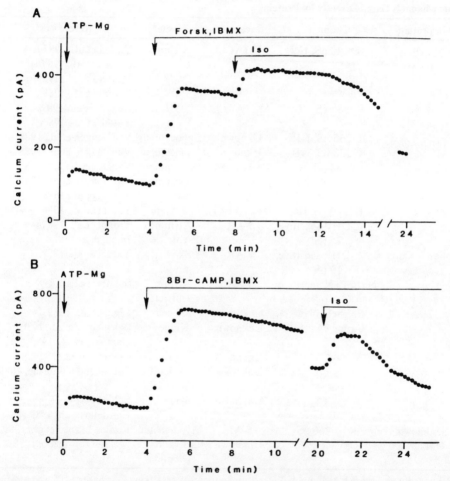

Figure 3. Time course of whole-cell Ca^{2+} currents recorded from single adult guinea pig ventricular myocytes. (*A*) Forskolin at 10 μM and IBMX at 100 μM were applied to the bath, then additional ISO a 1 μM was superfused when indicated. (*B*) 8-Br-cAMP at 58 μM and IBMX at 100 μM were applied, then additional ISO at 1 μM was superfused when indicated. In *A* and *B*, cells were dialyzed with ATP-Mg at 2 mM and Ca^{2+} currents were measured by applying a 100-ms clamp pulse to 0 mV from a holding potential of −50 mV at 0.05 Hz. *A* and *B* are separate experiments. Cells were placed in a recording solution containing: 2 mM $CaCl_2$, 135 mM tetraethylammonium Cl, 5 mM 4-aminopyridine, 1 mM $MgCl_2$, 10 mM glucose, 10 mM HEPES, and 0.01 mM tetrodotoxin (pH 7.3). The pipette solution was 110 mM Cs-aspartate, 20 mM $CsCl_2$, 2 mM ATP, 2 mM $MgCl_2$, 0.1 GTP, 5 mM EGTA, and 5 mM HEPES (pH 7.3).

satisfactory aspect of the fast response is that it allows an explanation for the beat-to-beat regulation of heart rate by cardiac sympathetic nerves that the indirect cytoplasmic pathway could not provide.

Conclusions

A single G protein may have several effectors, and activation by a specific receptor agonist can produce a coordinated response at the membrane level itself. Ionic channels as targets for G protein membrane messengers provide immediate global changes in membrane resting or action potentials which may then alter cellular function quickly. In addition, cytoplasmic messengers are activated indirectly and these produce complex cellular responses that have slower time courses. Ionic channels are also targets for these indirect G protein effects. Therefore, we view the direct interaction between G proteins and ionic channels as a membrane network that can be switched on and off rapidly but which may, through cytoplasmic messengers, produce slower longer-lasting effects.

Acknowledgments

This work was supported in part by grants NS-23877, HL-36930, HL-37044, and HL-39262 from the National Institutes of Health to A. M. Brown.

References

Ashkenazi, A., J. W. Winslow, E. G. Peralta, and G. L. Peterson. 1987. An M2 muscarinic receptor subtype coupled to both adenylyl cyclase and phosphoinositide turnover. *Science*. 238:272–275.

Birnbaumer, L., J. Codina, R. Mattera, A. Yatani, N. Scherer, M.-J. Toro, and A. M. Brown. 1987. Signal transduction by G proteins. *Kidney International*. 32:S14–S37.

Breitweiser, G. E., and G. Szabo. 1985. Uncoupling of cardiac muscarinic and β-adrenergic receptors from ion channels by a guanine nucleotide analogue. *Nature*. 317:538–540.

Brown, A. M., and L. Birnbaumer. 1988. Direct G protein gating of ion channels. *American Journal of Physiology*. 23:H410–H410.

Cantiello, H. F., C. R. Patenaude, and D. A. Ausiello. 1989. GTP-binding protein regulates the activity of Na^+ channels from the epithelial cell line A6. *Journal of General Physiology*. 94:3a. (Abstr.)

Codina, J., G. Grenet, A. Yatani, L. Birnbaumer, and A. M. Brown. 1987a. Hormonal regulation of pituitary GH_3 cell K^+ channels by G_k is mediated by its alpha subunit. *FEBS Letters*. 216:104–106.

Codina, J., A. Yatani, D. Grenet, A. M. Brown, and L. Birnbaumer. 1987b. The alpha subunit of G_k opens atrial potassium channels. *Science*. 236:442–445.

DiFrancesco, D., and C. Tromba. 1988. Muscarinic control of the hyperpolarization-activated current (i, f,) in rabbit sino-atrial node myocytes. *Journal of Physiology*. 405:493–510.

Dixon, R. A. F., B. K. Kobilka, D. J. Strader, J. L. Benovic, H. C. Dohlman, T. Frielle, M. A. Bolanowski, C. D. Bennett, E. Rands, R. E. Diehl, R. A. Mumford, E. E. Slater, I. S. Sigal, M. G. Caron, R. J. Lefkowitz, and C. D. Strader. 1986. Cloning of the gene and cDNA for mammalian beta-adrenergic receptor and homology with rhodopsin. *Nature*. 321:75–79.

Dunlap, K., G. G. Holz, and S. G. Rane. 1987. G proteins as regulators of ion channel function. *Trends in Neuroscience.* 10:241–244.

Eddlestone, G. T., B. Ribalet, and S. Ciani. 1989. Differences in K channel behavior in glucose-responsive and glucose-unresponsive insulin-secreting tumor cell lines. *Biophysical Journal.* 55:541a. (Abstr.)

Giles, W. R., and Y. Imaizumi. 1988. Comparison of potassium currents in rabbit atrial and ventricular cells. *Journal of Physiology.* 405:123–145.

Gilman, A. G. 1987. G proteins: transducers of receptor-generated signals. *Annual Review of Biochemistry.* 56:615–649.

Hisatome, I., T. Kiyosue, S. Imanishi, and M. Arita. 1985. Isoproterenol inhibits residual fast channel via stimulation of beta-adrenoreceptors in guinea-pig ventricular muscle. *Journal of Molecular and Cellular Cardiology.* 17:657–665.

Imoto, Y., A. Yatani, J. P. Reeves, J. Codina, L. Birnbaumer, and A. M. Brown. 1988. The α subunit of G_s directly activates cardiac calcium channels in lipid bilayers. *American Journal of Physiology.* 255:H722–H728.

Kameyama, M., F. Hoffman, and W. Trautwein. 1985. On the mechanism of beta-adrenergic regulation of the Ca channel in the guinea-pig heart. *Pflügers Archiv.* 405:285–293.

Kim, D., D. L. Lewis, L. Graziadei, E. J. Neer, D. Bar-Sagi, and D. E. Clapham. 1989. G-protein $\beta\gamma$-subunits activate the cardiac muscarinic K^+ channel via phospholipase A_2. *Nature.* 337:557–560.

Kurachi, Y., H. Ito, T. Sugimoto, T. Shimizu, I. Mike, and M. Ui. 1989. Arachidonic acid metabolites as intracellular modulators of the G protein-gated cardiac K^+ channel. *Nature.* 337:555–557.

Lacerda, A. E., D. Rampe, and A. M. Brown. 1988. Effects of protein kinase C activators on cardiac Ca^{2+} channels. *Nature.* 335:249–251.

Light, D. B., D. Ausiello, and B. A. Stanton. 1989. Guanine nucleotide binding protein, α_i^*-3, directly activates a cation channel in rat-renal inner medullary callecting duct cells. *Journal of Clinical Investigation.* 84:352–356.

Ma, J., and R. Coronado. 1988. Heterogeneity of conductance states in calcium channels of skeletal muscle. *Biophysical Journal.* 53:387–395.

Maguire, M. E., and J. J. Erdos. 1980. Inhibition of magnesium uptake by beta-adrenergic agonists and protstaglandin E1 is not mediated by cyclic AMP. *Journal of Biological Chemistry.* 255:1030–1035.

Mattera, R., M. P. Graziano, A. Yatani, Z. Zhou, R. Graf, J. Codina, L. Birnbaumer, A. G. Gilman, and A. M. Brown. 1989a. Splice variants of the α subunit of the G protein G_s activate both adenylyl cyclase and calcium channels. *Science.* 243:804–807.

Mattera, R., A. Yatani, G. E. Kirsch, R. Graf, J. Olate, J. Codina, A. M. Brown, and L. Birnbaumer. 1989. Recombinant $\alpha_i - 3$ subunit of G protein activates G_k-gated K^+ channels. *Journal of Biological Chemistry.* 264:465–471.

Parent, L., and R. Coronado. 1989. Reconstitution of the ATP-regulated potassium channel of skeletal muscle: activation by a G protein-dependent process. *Journal of General Physiology.* 94:445–463.

Ramos-Franco, J., L. Toro, and E. Stefani. 1989. GTPγS enhances the open probability of K_{Ca} channels from myometrium incorporated into bilayers. *Biophysical Journal.* 55:536a. (Abstr.)

Schubert, B., A. M. J. VanDongen G. E. Kirsch, and A. M. Brown. 1989a. β-adrenergic inhibition of cardiac sodium channels by dual G protein pathways. *Science*. 245:516–519.

Schubert, B., A. M. J. VanDongen, G. E. Kirsch, and A. M. Brown. 1989b. Modulation of cardiac Na channels by β-adrenoreceptors and the G protein, G_s. *Biophysical Journal*. 55:229a. (Abstr.)

Schwiebert, E. M., D. B. Light, and B. A. Stanton. 1989. A G protein, G_{i-3}, regulates a chloride channel in renal cortical collecting duct cells. *Journal of General Physiology*. 94:6a. (Abstr.)

Stryer, L. 1986. Cyclic GMP cascade of vision. *Annual Review of Neuroscience*. 9:87–119.

Trautwein, W., and A. Cavalie. 1985. Cardiac calcium channels and their control by neurotransmitters and drugs. *Journal of the American College of Cardiology*. 6:1409–1416.

VanDongen, T., J. Codina, J. Olate, R. Mattera, R. Joho, L. Birnbaumer, and Brown. A. M. 1988. Newly identified brain potassium channels gated by the guanine nucleotide binding protein G_o. *Science*. 242:1433–1437.

Walsh, K. B., and R. S. Kass. 1988. Regulation of a heart potassium channel by protein kinase A and C. *Science*. 242:67–69.

Yanagihara, K., and H. Irisawa. 1980. Inward current activated during hyperpolarization in the rabbit sinoatrial node cell. *Pflügers Archiv*. 385:11–19.

Yatani, A., and A. M. Brown. 1989. Rapid β-adrenergic modulation of cardiac calcium channel currents by a fast G protein pathway. *Science*. 245:71–74.

Yatani, A., J. Codina, A. M. Brown, and L. Birnbaumer. 1987a. Direct activation of mammalian atrial muscarinic potassium channels by GTP regulatory protein, G_k. *Science*. 1235:207–211.

Yatani, A., J. Codina, Y. Imoto, J. P. Reeves, L. Birnbaumer, and A. M. Brown. 1987b. A G protein directly regulates mammalian cardiac calcium channels. *Science*. 238:1288–1292.

Yatani, A., J. Codina, R. D. Sekura, L. Birnbaumer, and A. M. Brown. 1987c. Reconstitution of somatostatin and muscarinic receptor mediated stimulation of K^+ channels by isolated G_k protein in clonal rat anterior pituitary cell membranes. *Molecular Endocrinology*. 1:283–289.

Yatani, A., Y. Imoto, J. Codina, S. L. Hamilton, A. M. Brown, and L. Birnbaumer. 1988a. The stimulatory G protein of adenylyl cyclase, G_s, also stimulates dihydropyridine-sensitive Ca^{2+} channels. *Journal of Biological Chemistry*. 263:9887–9895.

Yatani, A., R. Mattera, J. Codina, R. Graf, K. Okabe, E. Padrell, R. Iyengar, A. M. Brown, and L. Birnbaumer. 1988b. The G protein-gated atrial K^+ channel is stimulated by three distinct $G_i\alpha$-subunits. *Nature*. 336:680–682.

Chapter 2

Modulation of Neuronal Calcium Currents by G Protein Activation

A. C. Dolphin and R. H. Scott

Department of Pharmacology, St. George's Hospital Medical School, London SW17 ORE, United Kingdom

Introduction

Neurotransmitters can affect the activity of an ion channel in several ways. First, a ligand-gated channel can be opened, and in this case the receptor forms part of the ion channel. Second, the opening of an ion channel can be modified either by its interaction with the α subunit of a G protein, or with a second messenger formed by interaction of an activated G protein with an enzyme. Indirect mechanisms will be slower in onset and offset than ligand-gated processes, which cease when the neurotransmitter is lost from the receptor. The termination of G protein-activated processes requires GTP hydrolysis and reassociation of the subunits (for reviews see Gilman, 1987; Dolphin, 1987), and second messenger–mediated processes also require metabolism of the second messenger and possibly dephosphorylation. Thus they are highly dependent on intracellular energy levels and temperature. One of the classes of ion channel whose activity is modified by such indirect mechanisms is the voltage-activated Ca channel.

Voltage-activated Ca channels are present throughout nerve cells (Llinas and Yarom, 1981; Tsien et al., 1988). They are involved in many of the activities of the neuron, including the initial response to neurotransmitters. This response may involve Ca channels indirectly since they will be activated after depolarization of the membrane by other transmitters such as glutamate (Choi, 1988). The activation of Ca channels may lead to the propagation of Ca spikes, which are involved in dendritic integration (Llinas and Yarom, 1981). They also play an essential role in the release of neurotransmitter, which depends critically on an influx of Ca^{2+} through voltage-gated channels at the presynaptic terminal (Augustine et al., 1987).

The different types of Ca channel that have been identified, and probably more that have not yet been clearly defined are certain, from their distinctive properties, to play quite individual roles in these specialized neuronal functions. For this reason, the mechanisms for modulation of Ca channels in neurons are likely to be more diverse, and to have a greater variety of consequences than in other cell types.

Types of Ca Channel in Neurons

The existence of the dihydropyridine-sensitive high voltage activated L channel in several neuronal cell classes has now been well documented (Carbone and Lux, 1984; Nowycky et al., 1985). The presence of a smaller conductance, low threshold or T channel was subsequently shown in peripheral neurons (Carbone and Lux, 1984). T channels have been reported to be blocked by octanol (Llinas, 1988), amiloride, and phenytoin (for review see Tsien et al., 1988), although these compounds all have other effects in the nervous system. T channels are also blocked in thalamic neurons by several other anticonvulsants (Coulter et al., 1989). It is likely that, because of its threshold for activation near the resting potential, this channel takes part in an important homeostatic function of neurons, regulating their rate of repetitive firing (Llinas and Yarom, 1981; Burlhis and Aghajanian, 1987; Coulter et al., 1989).

Another distinct subtype of Ca channel appears to be present only in neurons. It was originally described in dorsal root ganglion neurons (DRGs), and has been termed an N channel. It is not dihydropyridine sensitive, and has a single-channel conductance and kinetics of activation and inactivation intermediate between T and L channels (Nowycky et al., 1985; Plummer et al., 1989). It has been suggested that activation of N channels underlies the transient component of the high threshold whole-cell current, with L channels being responsible for the more sustained component. However, the relative contribution of N and L channels to the whole-cell current in neurons is not clear-cut (Swandulla and Armstrong, 1988). The problem arises because of their similar threshold for activation and the lack of a specific antagonist for N channels. In addition, although the high threshold current due to activation of N channels is thought to inactivate quite rapidly in DRGs; in sympathetic neurons N channels are less easily distinguished from L channels by their rate of inactivation (Tsien et al., 1988).

Modulation of High Threshold Calcium Channels by G Protein Activation

Many investigations have studied the interaction between Ca currents and GTP-binding proteins. In most cases, the unequivocal demonstration of a direct interaction has not been achieved but in the following section the possibility of direct coupling of Ca channels to the different classes of G proteins will be assessed.

In many cell types such as DRGs, which exhibit Ca^{2+} action potentials, or in which there is a substantial calcium-dependent plateau phase of the action potential, neurotransmitters and neuromodulators have been observed to decrease this component (Dunlap and Fischbach, 1981). An example is shown in Fig. 1 A of the reduction of DRG action potential duration by the $GABA_B$ agonist (−) − baclofen. Subsequently it has been observed that calcium currents in many cell types are inhibited by a variety of hormones and neurotransmitters. In Fig. 1 B, a current-voltage relationship is shown for baclofen (100 μM) inhibition of Ca channel currents in cultured rat DRGs.

The initial evidence that a GTP-binding protein is involved was suggested by the finding that the inhibition of Ca currents by GABA, baclofen, and noradrenaline in chick and rat DRG neurons could be enhanced by GTP analogues (Scott and Dolphin, 1986), and inhibited by GDP analogues (Holz et al., 1986), and by pertussis toxin. There are now many other examples in other types of neurons and neuronal cell lines (for review see Tsien et al., 1988). In many cases neurotransmitter modulation of Ca channel currents involves selective inhibition of the transient component of the high threshold current. An example is shown in Fig. 1 C, in which 50 μM baclofen differentially inhibits the transient component of the high threshold Ca channel current recorded in cultured rat DRGs. This has been suggested to represent inhibition of N current (for review see Tsien et al. 1988). In some cases this interpretation must be regarded with caution, particularly where Ca^{2+} is the charge carrier, because of the additional complication of Ca^{s+}-dependent inactivation (Eckert and Tillotson, 1981). In some studies it is clear that neruotransmitters have a complex effect on several components of the current; for example, a diversity of responses to noradrenaline has been observed in NG 108-15 cells (Docherty and McFadzean, 1989). Indeed, in DRG cells held at a depolarized holding potential of −30 mV, baclofen was still able to inhibit the remaining current (Fig. 1 D) even though only noninactivating current remained, since all the transient high threshold current was inactivated at this holding potential (Dolphin and Scott, 1990).

GTP analogues have a marked effect on Ca channel currents. They clearly inhibit the transient component of the whole-cell current in rat DRGs, and they slow current activation (Fig. 2 A; Dolphin and Scott, 1987; Dolphin et al., 1988). This is shown by photo-release of the GTP analogue GTP-γ-S inside the cell, which results in gradual inhibition of the calcium current, the transient current being completely abolished, whereas the sustained current present at the end of the step is partially reduced (Fig. 2 A). However, this cannot necessarily be taken as evidence of inhibition of an N current, since it has also been observed in AtT-20 cells which do not possess N channels (Lewis et al., 1986). Our studies using photo-release of "caged" GTP analogues show that the calcium channel current that remains available at −30 mV (comprising a largely dihydropyridine-sensitive current) can also be partly inhibited by G protein

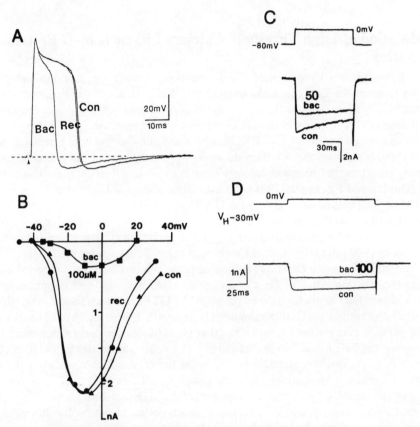

Figure 1. The effect of (−)-baclofen on DRG action potentials and calcium channel currents. (A) Reduction by 100 μM (−)-baclofen of the action potential duration, with no effect on resting membrane potential. The action potential was evoked by a brief depolarizing pulse (*arrowhead*), and was prolonged by 2.5 mM tetraethylammonium (TEA). (B) Current-voltage relationship, showing inhibition by 100 μM baclofen (■) of control calcium channel current (▲), with complete recovery at 5 min (●). The current was carried by Ba^{2+} (2.5 mM). All currents are leak-subtracted. V_H − 80 mV. (C) Differential inhibiton of the transient I_{Ba} by 50 μM baclofen. (D) Inhibition by 100 μM baclofen of the sustained current available from −30 mV.

activation in DRGs (Fig. 2 B; Scott and Dolphin, unpublished results). This conclusion is further supported by the ability of baclofen to inhibit the sustained Ca channel current remaining in the presence of GTP-γ-S (Fig. 2 C). Thus, in our hands, GTP-γ-S and the agonist baclofen inhibit both the high threshold transient current, in agreement with Plummer et al. (1989) and the high threshold sustained current, in agreement with Holz et al. (1986). The extent of Ca current inhibition by neurotransmitters in different cell types must also depend on the receptor–G protein channel stoichiometry and to what extent these elements are able to move within the plane of the membrane.

The finding that exogenously added GTP analogues (GTP-γ-S, GppNHp) inhibit calcium channel currents in many neuronal cells (DRGs, sympathetic neurons, cortical neurons, and cerebellar granule cells) led to our suggestion that endogenous GTP itself might modulate calcium currents (Dolphin and Scott, 1987). In the presence of internal GDP-β-S, which competes with endogenous GTP for binding to the G protein, the transient component of calcium currents is enhanced in amplitude (Fig. 3, *left*) (Dolphin and Scott, 1987). This is also the case in pertussis toxin–treated cells, in which activation of G_i/G_o is prevented by its ADP-ribosylation (Fig. 3, *left*). In

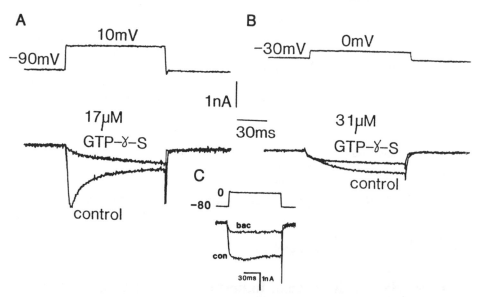

Figure 2. The effect of internal GTP-γ-S on calcium channel currents. (*A*) The larger trace shows the maximum I_{Ba} recorded from V_H −80 mV in the presence of 100 μM caged GTP-γ-S. It was stable for 5 min; subsequently, three flashes were given to release GTP-γ-S (17 μM), and the smaller current was recorded after 5 min. It was slowly activating and sustained. (*B*) The larger trace shows the maximum I_{Ba} recorder in another cell from V_H −30 mV. After six flashes to release 31 μM GTP-γ-S, the smaller trace was recorded after 5 min. (*C*) (−)-Baclofen (50 μM) is able to inhibit the sustained I_{Ba} recorded from V_H −80 mV in the presence of 500 μM internal GTP-γ-S in the patch pipette solution.

addition, high concentrations of GTP (1 mM) have a similar although not such a marked effect as GTP-γ-S on the calcium channel current (Fig. 3, *right*) (Dolphin and Scott, 1989). These findings suggest that there is a tonic partial inhibition of calcium channel currents in DRGs, by a proportion of activated G proteins.

Voltage Sensitivity of Calcium Channel Current Inhibition

It has recently been observed by Bean (1989) that noradrenaline did not inhibit frog DRG calcium channel currents activated by large depolarizations so that the current flowing was outward, due to Cs^+ passing out of the cell through Ca channels. In

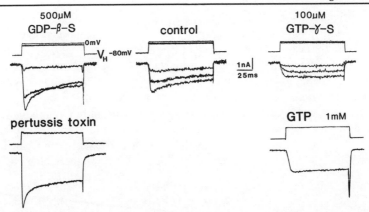

Figure 3. The effect of guanine nucleotide analogues and pertussis toxin on calcium channel currents. A family of control calcium channel currents are shown in the center panel. For comparison, in the left-hand panel two treatments (internal GDP-β-S or pretreatment with pertussis toxin), which reduce G protein activation by endogenous GTP, increase the transient component of I_{Ba}. In the right-hand panel GTP-γ-S and GTP itself reduced the transient component of I_{Ba}.

contrast, we have observed that outward Ca channel current activated by step depolarizations to +120 mV is smaller in amplitude in the presence of internal GTP-γ-S (Dolphin and Scott, 1990) (Fig. 4). It is also inhibited by (−)-baclofen. However, whereas the inward current activation is very markedly slowed by agonist and G protein activation, this effect is much reduced for the outward current. Our finding suggests that G protein modulation of Ca channel currents is able to persist during brief periods of extreme depolarization.

G Protein Species Involved in Inhibition of Ca Channel Currents

The G protein involved in inhibition of Ca channel current is clearly pertussis toxin sensitive (Holz et al., 1986; Dolphin and Scott, 1987). Identification of the G protein involved in coupling receptors to Ca current inhibition has been aided by experiments in which G proteins are included in the patch pipette when recording Ca currents in pertussis toxin–treated cells. It has been found that exogenously added G proteins can restore the ability of neurotransmitters to inhibit Ca currents. These experiments have generally suggested that G_o or its α subunit is more effective than G_i in restoring

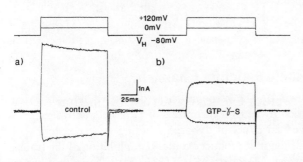

Figure 4. The effect of GTP-γ-S on inward and outward calcium channel currents. Currents were recorded from V_H −80 mV in a control cell (a) and in a cell recorded with a patch pipette containing 100 μM GTP-γ-S (b). The maximum inward current was recorded at 0 mV and the outward current at +120 mV.

coupling (Hescheler et al., 1987; Ewald et al., 1988). In several systems, the use of anti–G protein antibodies which inhibit function have confirmed that a G_o protein mediates inhibition of neuronal Ca currents by neurotransmitters (Ewald et al., 1988; Harris-Warwick et al., 1988; Brown et al., 1989).

Second Messenger Involvement in the Inhibition of Calcium Currents

There is no unequivocal evidence that the inhibition of Ca channels by neurotransmitter occurs by direct interaction with activated G protein subunits. However, most evidence suggests that neither activation nor inhibition of adenylate cyclase is involved in the response (McFadzean and Docherty, 1989; Dolphin et al., 1989a). Nevertheless the activity of L channels (in neurons as in heart) is increased by cyclic AMP–dependent phosphorylation (Gray and Johnston, 1987; Lipscombe et al., 1988; Dolphin et al., 1989a).

The evidence concerning the obligatory mediation of protein kinase C (PKC) in the inhibition of Ca^{2+} currents by neurotransmitters remains equivocal. Rane and Dunlap (1986) initially demonstrated an inhibition of the sustained Ca^{2+} current in chick DRGs by phorbol esters and oleoylacetylglycerol (OAG). However, we observed no inhibitory effect of dioctanoyl glycerol on I_{Ba} (Fig. 5 C). Several studies have failed to show either an effect on Ca^{2+} currents of PKC activators or the ability of inhibitors of PKC to prevent the response of calcium currents to neurotransmitters or GTP analogues (Brezina et al., 1987; Wanke et al., 1987; Dolphin et al., 1989a; McFadzean and Docherty, 1989). The PKC inhibitors polymixin B and H7 both produced some inhibition of calcium channel currents, but did not prevent the effects of guanine nucleotide analogues (Dolphin et al., 1989a). A recent study by Hockberger et al. (1989) has shown that although OAG inhibits DRG Ca^{2+} currents when applied externally, it had no effect when included intracellularly. This indicates that a direct membrane effect may also contribute to the response to PKC activators.

The neurotransmitters and neuromodulators that inhibit calcium currents are not normally considered to be in the class of "calcium mobilizing" agonists that activate phospholipase C (PLC). However, we have found $(-)$-baclofen (100 μM) to increase the production of total inositol phosphates in DRG neurons by 22% in 30 s (Dolphin et al., 1989a). For comparison, the enhancement by bradykinin (1 μM) was ~70% (Fig. 5 A). However, we observed bradykinin to have on average little effect on I_{Ba} or I_{Ca}. In some cells (7/16) it produced a marked increase of I_{Ba} by 72 ± 16% (Dolphin et al., 1989a; McGuirk and Dolphin, unpublished results (e.g., Fig. 5 B). In contrast, as previously shown, baclofen produces a substantial inhibition. This suggests that activation of PLC is not required for inhibition of calcium channel currents by baclofen.

Inhibition of Low Threshold Calcium Channels

There are contradictory findings on the responsiveness of low voltage–activated T currents to inhibitory modulation by neurotransmitters or G protein activation. Gross and MacDonald (1987) showed no effect of dynorphin acting at κ receptors on low threshold currents in DRGs, although it produced a marked inhibition of the transient high threshold current. In contrast, Marchetti et al. (1986) have shown that, whereas dopamine slows the rate of activation and reduces the amplitude of high threshold currents in DRG neurons, it inhibits low threshold currents with no change in their kinetics. Low threshold single-channel activity recorded from inside-out patches was also inhibited by dopamine. Similarly we have shown (Scott and Dolphin, 1987a) an adenosine agonist, 2-chloroadenosine, to inhibit low threshold Ca^{2+} currents in rat DRGs.

Further evidence in these cells for the association of T channels with a GTP-binding protein comes from our studies showing that photo-release of 10–20 μM GTP-γ-S from its inactive "caged" precursor inhibits T currents in a pertussis

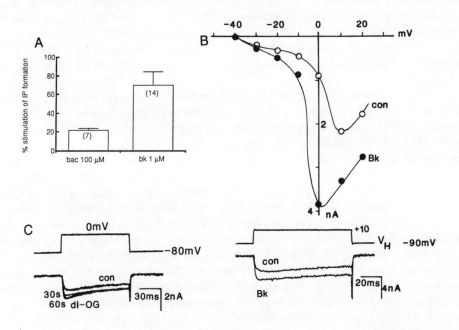

Effect of PLC and PKC activation on I_{Ba}

Figure 5. The effect of activation of PLC and PKC on calcium channel currents. (*A*) The stimulation of total inositol phosphate formation in 30 s by 100 μM (−)-baclofen, compared with 1 μM bradykinin. (*B*) In contrast to the inhibition of I_{Ba} by baclofen, 1 μM bradykinin either had no effect on I_{Ba} or produced a rapid increase in I_{Ba}, and shifted the voltage of activation to more negative potentials. The traces shown were recorded at +10 mV. (*C*) Dioctanoylglycerol (di-OG: 20 μM) either had no effect on I_{Ba} or produced a small enhancement over ~60 s.

toxin–sensitive manner (Dolphin et al., 1989b) (Fig. 6 *A*) Inhibition of 87 ± 9% was produced by liberation of 17 μM GTP-γ-S ($n = 4$).

A recent finding is that photo-release of a low concentration of GTP-γ-S (6 μM) consistently increases T current by 54 ± 20% ($n = 6$) (Fig. 6 *B*) (Dolphin et al., 1989b). The increase is sensitive neither to pertussis toxin nor to cholera toxin (Scott et al., in press). Baclofen also has dual effects on the low threshold Ca channel current: a low concentration (2 μM) increased the T current by 25 ± 4% ($n = 6$), whereas 100 μM produced 31 ± 7% inhibition ($n = 5$) (Scott et al., in press). The physiological relevance of neurotransmitter and G protein–mediated up- and down-modulation of

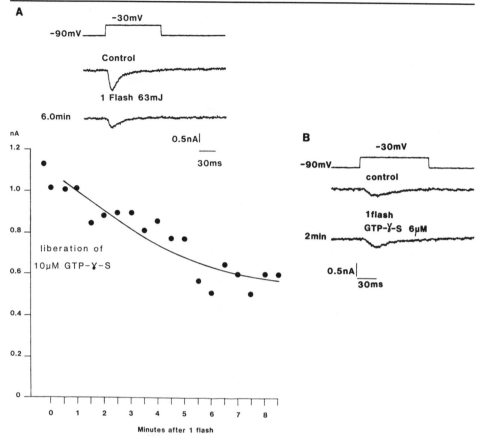

Figure 6. The effect of GTP-γ-S on T currents. (A) 100 μM caged GTP-γ-S was included in the patch pipette solution, and allowed to equilibrate with the cell for 10 min before giving a single flash, which photolyzed the caged compound with 10% efficiency, yielding ~10 μM free GTP-γ-S inside the cell. T currents were activated every 30 s at a clamp potential of −30 mV from a holding potential of −90 mV. The T current was stable before GTP-γ-S liberation. The peak amplitude of the T current is plotted against time after photo-release of GTP-γ-S. The traces show a control T current, and the current was recorded 6 min after liberation of 10 μM GTP-γ-S (B) Liberation of a lower concentration (6 μM) GTP-γ-S produced an increase in T current amplitude.

neuronal T currents remains to be determined, but these currents are clearly important in regulating patterns of neuronal firing (Llinas, 1988).

Interaction between Calcium Channel Ligand Binding Sites and G Protein Activation

Recent work has suggested an interaction between G protein activation and the effect of Ca^{2+} channel ligands on L-type Ca channels in cultured rat DRGs and sympathetic neurons (Scott and Dolphin, 1987b, 1988; Dolphin and Scott, 1989). Ca^{2+} channel antagonists, including nifedipine (5 μM), (−)−202-791 (0.5–5 μM), diltiazem (30 μM), and D_{600} (10 μM) were observed to show only agonist properties in the presence of internal guanine nucleotide analogues. For example, whereas in control cells 5 µM nifedipine produced a transient enhancement followed by a persistent inhibition of I_{Ba}, in the presence of internal GTP-γ-S, only enhancement was observed (Scott and Dolphin, 1987) and this was prolonged for the duration of the experiment. This result is not due to an effect on the steady-state inactivation of I_{Ba}, since nifedipine, whether acting as an agonist or an antagonist shifted the steady-state inactivation curve to more hyperpolarized potentials (Dolphin and Scott, 1989). Another example, using 0.5 µM (−)−202-791 is shown in Fig. 7 A.

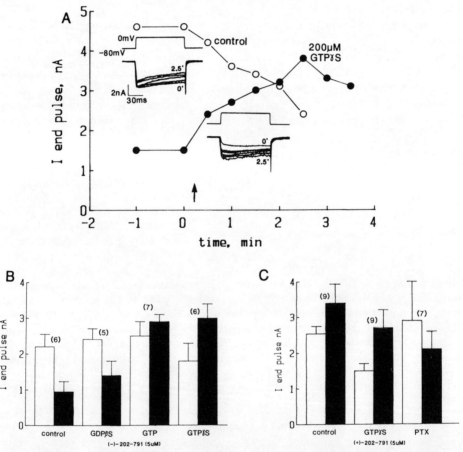

Figure 7. The effect of alterations in G protein activation on the responses of calcium channel currents to the (+) and (−) isomers of 202-791. (A) In a control cell (○) 0.5 µM (−)−202-791 inhibits I_{Ba}, whereas in a cell containing 200 µM GTP-γ-S (●) the same compound shows agonist properties. The application of (−)−202-791 was from a blunt pressure pipette placed near the cell. Its application was started at the time indicated by the arrow and continued for the rest of the experiment. (B) The effect of 5 µM (−)−202-791 is shown on the maximum I_{Ba} from V_H −80 mV in control cells, in the presence of 1 mM internal GDP-β-S, 1 mM GTP, or 200 µM GTP-γ-S. The open bars represent the mean I_{Ba} (± SEM) measured at the end of the 100-ms step command. The black bars represent the maximum effect observed of (−)−202-791. (C) The effect of 5 µM (+)−202-791 is shown (as in B) in control cells, in the presence of GTP-γ-S, and in pertussis toxin–treated cells.

Further studies showed that agonist responses to both "antagonist" and "agonist" Ca^{2+} channel ligands were blocked by pertussis toxin in DRGs (Fig. 7, B and C). In addition, the rate of onset of the antagonist response to $(-)-202-791$ was increased by GDP-β-S (Dolphin and Scott, 1989). Similarly, internal GTP (1 mM) also induced agonist responses to $(-)-202-791$, although these were smaller in amplitude than those seen with GTP-γ-S (Fig. 7 B). As a control that GTP-γ-S had not fundamentally affected the pharmacology of the Ca channels, the ability of several other Ca channel blockers to inhibit control Ca channel currents and those in the presence of GTP-γ-S was compared. ω-conotoxin (1 μM; Fig. 8), ω-agatoxin IA (10 nM, Adams et al., 1989), and Cd^{2+} (100 μM) inhibited the Ca channel currents similarly under the two conditions.

These findings indicate that in the intact neuronal cells studied, a G protein that can be activated by GTP or its nonhydrolyzable analogues is an essential requirement for Ca^{2+} channel ligands to act as agonists. The proposed model is described in more detail in Dolphin and Scott (1989). Since purified dihydropyridine receptors have Ca^{2+} channel activity when inserted in lipid bilayers, and show an agonist response to Bay K 8644 (Hymel et al., 1988), one possible explanation is that inactive G proteins prevent the channel from existing in the conformation in which the ligands can act as agonists.

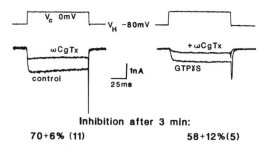

Figure 8. The effect of ω-conotoxin (ω-CgTx) on calcium channel currents. ω-CgTx (1 μM) inhibited I_{Ba} in control cells (*left*) and in the presence of internal GTP-γ-S (*right*). An initial rapid partial inhibition was followed by complete inhibition of I_{Ba} in some cells where application was continued for 5–10 min.

Under control conditions it is likely that there is partial tonic activation of some endogenous G proteins and that these are responsible for the initial transient agonist response observed to many calcium channel antagonists (Scott and Dolphin, 1987b). Although most binding studies have shown equivocal effects of guanine nucleotide analogues on calcium channel ligand binding, a recent report has found that GTP-γ-S enhanced the binding of Bay K 8644 to rat brain membranes (Bergamaschi et al., 1988). Clearly the interaction between calcium channels and G proteins is complex, and is likely to be unraveled by reconstitution experiments with pure receptors, channels, and G proteins.

Which Types of Ca Channels Mediate Transmitter Release?

The release of neurotransmitters from presynaptic terminals is dependent on an influx of Ca^{2+} through voltage-sensitive channels. The subtypes of Ca channel involved in transmitter release remain equivocal. It has been suggested that N channels are the most important (Hirning et al., 1988; Tsien et al., 1988) primarily because transmitter release is not markedly sensitive to dihydropyridine Ca channel antagonists, although it is sensitive to agonists (Rane et al., 1987). However, dihydropyridine antagonists show a very marked voltage sensitivity, and are poorly effective at hyperpolarized membrane potentials (Sanguinetti and Kass, 1984). When cultured cerebellar neurons are maintained in a depolarized state in 50 mM K^+, Ca^{2+}-free medium, and glutamate release is stimulated by Ca^{2+}, this release is almost completely inhibited by dihydropyridines, the IC_{50} for $(-)-202$-791 being 1 nM (Huston et al., in press) Thus dihydropyridine-sensitive Ca channels are able to subserve transmitter release.

The other agent that has been used to dissect out the channels involved in transmitter release is ω-conotoxin. This inhibits transmitter release in many systems, but there is disagreement as to whether it selectively inhibits N channels (Plummer et al., 1989), or whether L channels are also affected (McCleskey et al., 1987).

A further argument for the prime importance of N channels in the mediation of transmitter release is that "N current" can be inhibited by various neurotransmitters that also subserve presynaptic inhibition. However, as discussed above, there is also evidence that L current can be inhibited by neurotransmitters. In addition, in recent experiments we have shown that $(-)$-baclofen is capable of inhibiting neurotransmitter release evoked by Ca^{2+} from cerebellar neurons maintained in a depolarized state, where the Ca channels involved are entirely dihydropyridine sensitive (Huston et al., in press). Thus the involvement of different Ca channel types in physiological synaptic transmission remains an open question.

Role of G Protein Modulation of Ca Channels in the Regulation of Transmitter Release

Many neurotransmitters have been shown to act on presynaptic receptors to inhibit neurotransmitter release, and these effects can be examined both at the biochemical and at the electrophysiological level. Classical presynaptic inhibition results from activation of $GABA_A$ receptors, and activation of a Cl^- conductance. However, presynaptic inhibition by other neurotransmitters may involve either an enhancement of a K^+ conductance, reducing the ability of action potentials to invade the terminal, or inhibition of Ca^{2+} currents, directly reducing the influx of Ca^{2+} which is a prerequisite for transmitter release. Both these processes may involve pertussis toxin–sensitive G proteins (Nicoll, 1988; Tsien et al., 1988).

In biochemical studies of neurotransmitter release there are many examples of the ability of pertussis toxin to prevent presynaptic inhibition particularly in cultured cells but also in slices (Allgaier et al., 1985; Dolphin and Prestwich, 1985; Kawata and Nomura, 1987; Ulivi et al., 1988; Holz et al., 1989).

In contrast, there is less electrophysiological evidence that presynaptic inhibition of excitatory post-synaptic potentials (EPSPs) is prevented by pertussis toxin, although an early study showed that presynaptic inhibition of the twitch response in guinea pig ileum by opiates and the α_2 agonist clonidine was inhibited by pertussis toxin (Tucker, 1984). In addition, Colmers and Williams (1988) showed that the EPSP recorded in dorsal raphé neurons upon local stimulation of slices could be inhibited by ~20% by a 5-HT_1 agonist, and this response was prevented by prior intraventricular administration of pertussis toxin. In addition, pertussis toxin prevents the inhibitory effect of adenosine at mammalian motor nerve endings (Silinsky et al., 1989).

The contradictory data relate particularly to $GABA_B$-mediated events. Whereas pertussis toxin has been shown to block the baclofen-mediated inhibition of noradrenaline, substance P, and glutamate release (Ulivi et al., 1988; Holz et al., 1989; Huston et al., in press), several studies have shown no effect of pertussis toxin on baclofen-mediated presynaptic inhibition of excitatory and inhibitory postsynaptic potentials (Colmers and Williams, 1988; Dutar and Nicoll, 1988; Harrison, 1988).

There are several possible explanations for the discrepancies outlined above. It may be that there is a large excess of G proteins in synaptic boutons, and only a small proportion of these are required to mediate a presynaptic inhibitory effect on the EPSP resulting from a single action potential. Since pertussis toxin treatment rarely results in complete ADP-ribosylation of all its substrate G proteins, particularly when applied to intact tissues, it is possible that sufficient G proteins remained unaffected. Penetration of pertussis toxin into presynaptic terminals may be less efficient than into postsynaptic elements. Another reason for the ineffectiveness of pertussis toxin might be that a proportion of G proteins in presynaptic terminals are tonically GTP-activated, or for another reason not associated with $\beta\gamma$, and thus they are not substrates for pertussis toxin.

In addition, it is clearly possible that presynaptic inhibition also occurs at least in part by a mechanism involving neither inhibition of Ca^{2+} currents nor activation of potassium currents by pertussis toxin–sensitive G proteins. A recent electrophysiological study of the $GABA_B$-induced presynaptic inhibition of sensory afferents showed that it involved neither a reduction in the size of the action potential invading the terminal nor an increase in the number of invasion failures (Peng and Frank, 1989). This would be compatible with $GABA_B$ inhibition of presynaptic Ca channels. It will be of great interest to investigate the role of G proteins in this model for presynaptic inhibition.

Conclusion

In the preceding sections we have outlined the evidence for the modulation of Ca^{2+} channel activity by G proteins. When considering the modulation of synaptic transmission, it is clear that neurotransmitter-mediated activation of postsynaptic K^+ conductances serves to limit neuronal excitability (Nicoll, 1988). Presynaptic inhibition by the same family of neurotransmitters and modulators may result from both activation of K^+ currents and inhibition of Ca^{2+} currents. However, many questions remain to be answered concerning the mechanisms responsible for this form of inhibition.

References

Allgaier, C., T. J. Feuerstein, R. Jakisch, and G. Hertting. 1985. Islet activating protein (pertussis toxin) diminishes α_2 adrenoceptor-mediated effects on noradrenaline release. *Naunyn-Schmiedeberg's Archives of Pharmacology.* 331:235–239.

Adams, M. E., V. P. Bindokas, A. C. Dolphin, and R. H. Scott. 1989. The inhibition of Ca^{2+} channel currents in cultured neonatal rat dorsal root ganglion (DRG) neurones by ω-Agatoxin-1A (a funnel web spider toxin) *Journal of Physiology.* 418:34P.

Augustine, G. J., M. P. Charlton, and S. J. Smith. 1987. Calcium action in synaptic transmitter release. *Annual Review of Neuroscience.* 10:633-693.

Bean, B. P. 1989. Neurotransmitter inhibition of neuronal calcium currents by changes in channel voltage-dependence. *Nature.* 340:153–156.

Bergamaschi, S., S. Govoni, P. Cominetti, M. Parenti, and M. Trabucchi. 1988. Direct coupling of a G protein to dihydropyridine binding sites. *Biochemical and Biophysical Research Communications.* 56:1279–1286.

Brezina, V., R. Eckert, and C. Erxleben. 1987. Suppression of calcium current by an endogenous neuropeptide in neurones of *Aplysia californica. Journal of Physiology.* 388:565–595.

Brown, D. A., I. McFadzean, and G. Milligan. 1989. Antibodies to the GTP binding protein G_o attenuate the inhibition of the calcium current by noradrenaline in mouse neuroblastoma x rat glioma (NG 108-15) hybrid cells. *Journal of Physiology.* 415:20P.

Burlhis, T. M., and G. K. Aghajanian. 1987. Pacemaker potentials of serotonergic dorsal raphé neurones: contribution of a low threshold Ca^{2+} conductance. *Synapse.* 1:582–588.

Carbone, E., and H. D. Lux. 1984. A low voltage activated fully inactivating Ca channel in vertebrate sensory neurones. *Nature.* 310:501–502.

Choi, D. W. 1988. Calcium-mediated neurotoxicity: relationship to specific channel types and role in ischaemic damage. *Trends in Neurosciences.* 11:465–469.

Colmers, W. F., and J. T. Williams. 1988. Pertussis toxin pretreatment discriminates between pre- and postsynaptic actions of baclofen in rat dorsal raphé nucleus in vitro. *Neuroscience Letters.* 93:300–306.

Coulter, D. A., J. R. Huguenard, and D. A. Prince. 1989. Specific petit mal anticonvulsants reduce calcium currents in thalamic neurons. *Neuroscience Letters.* 98:74–78.

Docherty, R. J., and I. McFadzean. 1989. Noradrenaline-induced inhibition of voltage-sensitive calcium currents in NG108-15 hybrid cells. *European Journal of Neuroscience.* 1:135–140.

Dolphin, A. C 1987. Nucleotide binding proteins in signal transduction and disease. *Trends in Neurosciences.* 10:53–57.

Dolphin, A. C., S. M. McGuirk, and R. H. Scott. 1989a. An investigation into the mechanisms of inhibition of calcium channel currents in cultured rat sensory neurones by guanine nucleotide analogues and (−)-baclofen. *British Journal of Pharmacology.* 97:263–273.

Dolphin, A. C., R. II. Scott, and J. F. Wootton. 1989b. Photorelease of GTP-γ-S inhibits a low threshold calcium channel current in cultured rat dorsal root ganglion (DRG) neurones. *Journal of Physiology.* 410:16P.

Dolphin, A. C., and S. A Prestwich. 1985. Pertussis toxin reverses adenosine inhibition of neuronal glutamate release. *Nature.* 316:148–150.

Dolphin, A. C., and R. H. Scott. 1987. Calcium channel currents and their inhibition by (−)-baclofen in rat sensory neurons: modulation by guanine nucleotides. *Journal of Physiology.* 386:1–17.

Dolphin, A. C., and R. H. Scott. 1989. Interaction between calcium channel ligands and guanine nucleotides in cultured rat sensory and sympathetic neurons. *Journal of Physiology.* 413:271–288.

Dolphin, A. C., and R. H. Scott. 1990. Calcium channel currents activated by large depolarisations: effect of guanine nucleotides and (−)-baclofen. *European Journal of Neuroscience.* 2:104–108.

Dolphin, A. C., J. F. Wootton, R. H. Scott, and D. R. Trentham. 1988. Photoactivation of intracellular guanosine triphosphate analogues reduces the amplitude and slows the kinetics of voltage activated calcium channel currents in sensory neurons. *Pflügers Archiv.* 411:628–636.

Dunlap, K., and G. Fischbach. 1981. Neurotransmitters decrease the calcium conductance activated by depolarisation of embryonic chick sensory neurons. *Journal of Physiology.* 317:519–535.

Dutar, P., and R. A. Nicoll. 1988. Pre- and postsynaptic $GABA_B$ receptors have different pharmacological properties. *Neuron.* 1:585–591.

Eckert, R., and D. L. Tillotson. 1981. Calcium-mediated inactivation of the calcium conductance in caesium-loaded giant neurones of *Aplysia california. Journal of Physiology.* 314:265–280.

Ewald, D. A., P. C. Sternweis, and R. J. Miller. 1988. Guanine nucleotide binding protein G_o induced coupling of neuropeptide Y receptors to Ca^{2+} channels in sensory neurons. *Proceedings of the National Academy of Sciences.* 85:3633–3637.

Gilman, A. G. 1987. G proteins: transducers of receptor generated signals. *Annual Review of Biochemistry.* 56:615–649.

Gray, R., and D. Johnston. 1987. Noradrenaline and β-adrenoceptor agonists increase activity of voltage-dependent calcium channels in hippocampal neurons. *Nature.* 327:620–622.

Gross, R. A., and R. L. MacDonald. 1987. Dynorphin A selectively reduces a large transient (N type) calcium current of mouse dorsal root ganglion neurons in cell culture. *Proceedings of the National Academy of Sciences.* 84:5469–5473.

Harrison, N. 1988. Baclofen decreases synaptic inhibition in cultured hippocampal neurons by a mechanism that is insensitive to pertussis toxin. *Society for Neuroscience Abstracts.* 14:439.9. (Abstr.)

Harris-Warwick, R. M., C. Hammond, D. Paupardin-Tritsch, V. Homburger, B. Rouot, J. Bockaert, and H. M. Gerschenfeld. 1988. An α_{40} subunit of a GTP binding protein immunologically related to G_o mediates a dopamine-induced decrease of a Ca^{2+} current in snail neurons. *Neuron.* 1:27–32.

Hescheler, J., W. Rosenthal, W. Trautwein, and G. Schultz. 1987. The GTP binding protein G_o regulates neuronal calcium channels. *Nature.* 325:445–447.

Hirning, D., A. P. Fox, E. W. McCleskey, B. M. Olivera, S. A. Thayer, R. J. Miller, and R. W. Tsien. 1988. Dominant role of N-type Ca^{2+} channels in evoked release of norepinephrine from sympathetic neurons. *Science.* 239:57–61.

Hockberger, P., M. Toselli, D. Swandulla, and H. D. Lux. 1989. A diacylglycerol analogue reduced neuronal calcium currents independently of protein kinase C activation. *Nature.* 338:340–342.

Holz, G. G., R. M. Kream, A. Spiegel, and K. Dunlap. 1989. G proteins couple α-adrenergic and $GABA_B$ receptors to inhibition of peptide secretion from peripheral sensory neurons. *Journal of Neuroscience.* 9:657–666.

Holz, G. G., S. G. Rane, and K. Dunlap. 1986. GTP binding proteins mediate transmitter inhibition of voltage dependent calcium channels. *Nature.* 319:670–672.

Huston, E., R. H. Scott, and A. C. Dolphin. 1990. A comparison of the effect of calcium channel ligands and $GABA_B$ antagonists on transmitter release and somatic calcium channel currents in cultured neurones. *Neuroscience.* In press.

Hymel, L., J. Streissnig, H. Glossmann, and H. Schindler. 1988. Purified skeletal muscle 1,4-dihydropyridine receptor forms phosphorylation dependent oligomeric calcium channels in planar bilayers. *Proceedings of the National Academy of Sciences.* 85:4290–4294.

Kawata, K., and Y. Nomura. 1987. Suppressing effect of pertussis toxin on clonidine induced inhibition of noradrenaline release from cerebral cortical slices of rats. *Neuroscience Research*. 4:236–240.

Lewis, D. L., F. F Weight, and A. Luini. 1986. A guanine nucleotide binding protein mediates the inhibition of voltage-dependent calcium current by somatostatin in a pituitary cell line. *Proceedings of the National Academy of Sciences*. 83:9035–9039.

Lipscombe, D., K. Bley, and R. W. Tsien. 1988. Modulation of neuronal Ca channels by cAMP and phorbol esters. *Society for Neuroscience Abstracts*. 14:64.12. (Abstr.)

Llinas, R. 1988. The intrinsic electrophysiological properties of mammalian neurons: insights into central nervous system function. *Science*. 242:1654–1664.

Llinas, R., and Y. Yarom. 1981. Properties and distribution of ionic conductances generating electroresponsiveness of mammalian inferior olivary neurons in vitro. *Journal of Physiology*. 315:569–584.

Marchetti, C., E. Carbone, and H. D. Lux. 1986. Effects of dopamine and noradrenaline on Ca channels of cultured sensory and sympathetic neurones of chick. *Pflügers Archiv*. 406:104–111.

McCleskey, E. W., A. P. Fox, D. Feldman, L. Y. Cruz, B. M. Olivera, R. W. Tsien, and D. Yoshikami. 1987. Direct and persistent blockade of specific types of calcium channels in neurons but not muscle. *Proceedings of the National Academy of Sciences*. 84:4327–4331.

McFadzean, I., and R. J. Docherty. 1989. Noradrenaline and enkephalin-induced inhibition of voltage-sensitive calcium currents in NG108-15 hybrids cells: transduction mechanisms. *European Journal of Neuroscience*. 1:141–147.

Nicoll, R. A. 1988. The coupling of neurotransmitter receptors to ion channels in the brain. *Science*. 241:545–551.

Nowycky, M. C., A. P. Fox, and R. W. Tsien. 1985. Three types of neuronal calcium channel with different calcium agonist sensitivity. *Nature*. 316:440–443.

Peng, Y., and E. Frank. 1989. Activation of $GABA_B$ receptors causes presynaptic inhibition at synapses between muscle spindle afferents and motoneurons in the spinal cord of bullfrogs. *Journal of Neuroscience*. 9:1502–1515.

Plummer, M. R., D. E. Logothetis, and P. Hess. 1989. Elementary properties and pharmacological sensitivities of calcium channels in mammalian peripheral neurons. *Neuron*. 1:1453–1463.

Rane, S. G., and K. Dunlap. 1986. Kinase C activator 1,2-oleoyl-acetylglycerol attenuates voltage-dependent calcium current in sensory neurons. *Proceedings of the National Academy of Sciences*. 83:184–188.

Rane, S. G., G. G. Holz, and K. Dunlap. 1987. Dihydropyridine inhibition of neuronal calcium current and substance P release. *Pflügers Archiv*. 409:361–366.

Sanguinetti, M. C., and R. S. Kass. 1984. Voltage-dependent modulation of Ca channel current in the calf cardiac Purkinje fiber by dihydropyridine calcium channel antagonists. *Circulation Research*. 55:336–348.

Scott, R. H., and A. C. Dolphin. 1986. Regulation of calcium currents by a GTP analogue: potentiation of (−)-baclofen mediated inhibition. *Neuroscience Letters*. 56:59–64.

Scott, R. H., and A. C. Dolphin. 1987*a*. Topics and Perspectives in Adenosine Research E. Gerlach, and B. F. Becker, editors. Springer-Verlag, Berlin. 549–558.

Scott, R. H., and A. C. Dolphin. 1987*b*. Activation of a G protein promotes agonist responses to calcium channel ligands. *Nature*. 330:760–762.

Scott, R. H., and A. C. Dolphin. 1988. The agonist effect of Bay K 8644 on neuronal calcium channel currents is promoted by G protein activation. *Neuroscience Letters.* 89:170–175.

Scott, R. H., J. F. Wootton, and A. C. Dolphin. 1990. Modulation of neuronal T type calcium channel currents by photoactivation of intracellular GTP-γ-S. *Neuroscience.* In press.

Silinsky, E. M., C. Solsona, and J. K. Hirsh. 1989. Pertussis toxin prevents the inhibitory effect of adenosine and unmasks adenosine-induced excitation of mammalian motor nerve endings. *British Journal of Pharmacology.* 97:16–18.

Swandulla, D., and C M. Armstrong. 1988. Fast deactivating calcium channels in chick sensory neurons. *Journal of General Physiology.* 92:197–218.

Tsien, R. W., D. Lipscombe, D. V. Madison, K. R. Bley, and A. P. Fox. 1988. Multiple types of neuronal calcium channels and their selective modulation. *Trends in Neurosciences.* 11:431–437.

Tucker, J. F. 1984. Effect of pertussis toxin on normorphine dependence and on acute inhibitory effects in the isolated guinea pig ileum. *British Journal of Pharmacology.* 83:326–328.

Ulivi, M., W. Wojcik, and E. Costa. 1988. Baclofen inhibits glutamic acid release from primary cerebellar granule cells through a pertussis toxin sensitive guanine nucleotide coupling protein. *Society for Neuroscience Abstracts.* 14:325.2. (Abstr.)

Wanke, E., A. Ferroni, A. Malgaroli, A. Ambrosini, T. Pozzan, and J. Meldolesi. 1987. Activation of a muscarinic receptor selectively inhibits a rapidly inactivated Ca^{2+} current in rat sympathetic neurons. *Proceedings of the National Academy of Sciences.* 84:4313–4317.

Chapter 3

A Comparison of the Roles of Purified G Protein Subunits in the Activation of the Cardiac Muscarinic K^+ Channel

Chaya Nanavati, David E. Clapham, Hiroyuki Ito, and Yoshihisa Kurachi

Departments of Pharmacology and of Physiology and Biophysics, Mayo Foundation, Rochester, Minnesota 55905; and the Second Department of Internal Medicine, Faculty of Medicine, University of Tokyo, Hongo, Bunkyo-ku, Tokyo 113, Japan

Introduction

The cardiac atrial muscarinic receptor is coupled to its effector, the inwardly rectifying, K$^+$-selective channel, $i_{K.ACh}$, via pertussis toxin (PTX)–sensitive guanine nucleotide–binding proteins (G proteins). The GTP dependence of this channel was demonstrated at the whole-cell level by Pfaffinger et al. (1985) and Breitwieser and Szabo (1985), and at the single-channel level by Kurachi et al. (1986a). These initial experiments have shown that acetylcholine (ACh) induces an inwardly rectifying K$^+$ current in whole-cell voltage-clamped atrial cells only when GTP was included in the patch pipette (Pfaffinger et al., 1985). In the presence of ACh, this current was irreversibly activated by intracellular application of the nonhydrolyzable GTP analogue, GppNHp (Breitwieser and Szabo, 1985). It could also be irreversibly activated in the absence of agonist, by the nonhydrolyzable GTP analogue, GTPγS (Kurachi et al., 1986b). At the single-channel level, when receptor was bound, channel activation

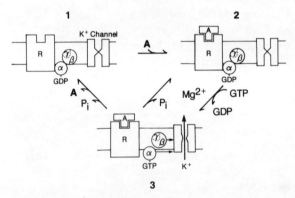

Figure 1. A simple model that incorporates the current views on the role of G proteins in signal transduction. State *1*: when the receptor is unoccupied, the G protein is in its inactive, holomeric, GDP-bound state. State *2*: when agonist, *A*, binds receptor, in the presence of intracellular GTP and Mg^{2+}, the GDP bound on the G$_\alpha$ subunit is exchanged for a GTP. State *3*: the G$_\alpha$-GTP and G$_{\beta\gamma}$ subunits are functionally dissociated, and can interact with effectors such as enzymes or channels. The arrows do not necessarily indicate direct activation of an effector by the G protein subunits. The G$_\alpha$ subunit with its intrinsic GTPase activity, hydrolyzes the GTP to yield a GDP-bound G$_\alpha$ subunit, and a free high-energy phosphate. The GDP-bound G$_\alpha$ and the G$_{\beta\gamma}$ can then return either to state *1* or to state *2*.

occurred only in the presence of intracellular GTP (Kurachi et al., 1986a). Adenosine (Kurachi et al., 1986b) and somatostatin (Lewis and Clapham, 1989) are also coupled to this channel via PTX-sensitive G proteins.

The heterotrimeric G proteins consist of α, β, and γ subunits. Fig. 1 is a schematic representation of the current model for G protein action. In the absence of agonist (state *1*), the G protein is in its inactive holomeric form with the G$_\alpha$ subunit bound to GDP. When agonist (*A*) binds the receptor (state *2*), in the presence of Mg^{2+} and GTP, the GDP bound on the G$_\alpha$ subunit is exchanged for GTP. The hydrophilic, membrane-anchored, activated G$_\alpha$-GTP and the hydrophobic G$_{\beta\gamma}$ subunits are functionally dissociated yielding two potential second messengers that can interact (either directly or indirectly) with effectors (state *3*). The G$_\alpha$ subunit hydrolyzes GTP to GDP, releasing a high-energy phosphate, and returning the G protein either to state *1* or *2*.

Both G$_\alpha$ (Codina et al., 1987; Cerbai et al., 1988; Logothetis et al., 1988; Kurachi et al., 1989) and $G_{\beta\gamma}$ subunits (Logothetis et al., 1987; Cerbai et al., 1988; Kurachi et al., 1989) can activate $i_{K.ACh}$. Activation by the $G_{\beta\gamma}$ subunit has been the source of considerable controversy since in most other G protein–linked systems studied, the G_α subunit interacts with the effector while the $G_{\beta\gamma}$ subunit acts as the modulator for activation by G_α (but see also Whiteway et al., 1989). Examples of such systems are β-adrenergic activation of adenylyl cyclase via the stimulatory G protein, G_s, and light activation of cGMP phosphodiesterase by the retinal G protein, G_{Tr} (transducin). It has been argued that activation of $i_{K.ACh}$ by $G_{\beta\gamma}$ was due to (*a*) the detergent, 3[(3-cholamidopropyl)di-methylammonio]-1-propanesulfonate (CHAPS), used to suspend the hydrophobic $G_{\beta\gamma}$ subunits, or (*b*) contamination by activated G_α. The results of experiments that address these criticisms have been presented elsewhere (Logothetis et al., 1988; Kurachi et al., 1989) and have been reviewed (Hartzell, 1988; Kurachi, 1989). Briefly, when applied alone, CHAPS, boiled $G_{\beta\gamma}$, $G_{Tr\beta\gamma}$, or nonactivated G_α fail to activate the channel while purified $G_{\beta\gamma}$ ($\leq$200 pM) and GTPγS-bound G_α ($\geq$10 pM) activate the channel in >95% and 20–50% of the patches, respectively. Furthermore, the activation by $G_{\beta\gamma}$ is not caused by G_α contamination since <0.01% nonactivated G_α contaminates the $G_{\beta\gamma}$ preparation (Logothetis et al., 1988). This paper presents more data which support a role for $G_{\beta\gamma}$ as an activator of this channel, and provides evidence suggesting that G_α may not be the sole physiological transducer between receptors and effectors.

Methods

Preparation of Atrial Cells

Atria from the hearts of 1–2-d-old rats were removed, finely minced, and dissociated enzymatically by treatment with trypsin and collagenase (Logothetis et al., 1988). Dissociated cells were suspended in DMEM (Gibco, Grand Island, NY) supplemented with 10% fetal calf serum, 20 U/ml penicillin, and 20 μg/ml streptomycin, plated on glass coverslips, incubated at 37°C in a 95% O_2/5% CO_2 atmosphere, and used within 8–24 h of culture (experiments for Figs. 3–5). Hearts from adult guinea pigs were perfused using a Langendorff procedure, with 0.04% wt/vol collagenase type I (Sigma Chemical Co., St. Louis, MO) in a nominally calcium-free Tyrode solution. Hearts were stored at 4°C in a high K/low Cl solution for later use. Single atrial cells were dissociated as described previously (Kurachi et al., 1986*b*) and used for the electrophysiology experiments of Figs. 6 and 7 *A*.

Electrophysiological Recordings

The inside-out mode of the patch-clamp technique (Hamill et al., 1981) was used to record single-channel currents with a List EPC-7 (Medical Systems, Corp., Great Neck, NY) patch-clamp amplifier. Single-channel activity was quantified on-line (described below) using an INDEC PDP-11/73 computer (Sunnyvale, CA). All experiments on neonatal rat atrial cells were conducted at room temperature (20–22°C) while those on adult guinea pig atrial cells were conducted at 35–37°C. The holding potential for all experiments was −80 mV. At this potential, the single-channel $i_{K.ACh}$ currents were 2.8 ± 0.12 pA. The bath and pipette solutions (called K-5) contained, in millimolar: 118.5 KCl, 21.5 KOH, 2 $MgCl_2$, 5 EGTA, 10 HEPES, pH 7.2. In some cases, 0.1–1 μM ACh was added to the pipette solution. Guanine nucleotides, G proteins, and detergents were either pipetted directly (10 µl added to a 90-µl recording chamber), or were perfused into the recording chamber. G_α subunits were dissolved in K-5 + 0.01% bovine serum albumin, while $G_{\beta\gamma}$ subunits which were made up in detergent were diluted into K-5.

One problem with the inside-out configuration was spontaneous seal-over of the patch, forming a full or partial vesicle and thus restricting subunit entry to the patch and making channels appear rounded or attenuated. The presence of a vesicle therefore could cause a false negative. At the end of every experiment, GTP or GTPγS was pipetted directly into the chamber. At the end of negative experiments, if the nucleotide (or analogue) failed to produce channel activation we assumed the patch had become a vesicle. Such patches were not included in the analysis.

Diffusion

G protein subunits added to the bath take a finite time to diffuse to the patch. Fig. 2 shows three physical configurations of the patch in the inside-out mode. In both the "ideal" and the "general" cases, the proteins have easy access to the patch interior. To calculate how long it would take for an added protein to access a partially sealed patch, we used a model based on empirically derived diffusion constants for whole-cell voltage-clamped cells (Pusch and Neher, 1988). For the hypothetical worst-case patch (2-µm patch diameter and a 10-nm access diameter), we calculated that a 42-kD

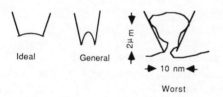

Figure 2. Three possible configurations for an inside-out patch relative to the patch pipette. In the "ideal" and "general" cases, applied proteins have easy access to the patch. However, in the contorted "worst" case, applied proteins will take 12 s to simply diffuse to the patch.

fluorescent dextran molecule would have a diffusion time constant of 12 s. Thus, for an extreme case, the patch concentration of a 42-kD protein should reach steady-state within ~3 time constants (~40 s). This analysis ignores the time necessary for the subunit to bind the channel effector site.

Quantitation of Channel Activity

The total current, I, flowing per second through N channels in a membrane patch is equal to $NP_o i$ where P_o is the probability of one channel being open and i is the single-channel current (2.8 pA for this channel). Since most of our patches contained more than one channel, we computed NP_o (N was unknown) as I/i. Filtered (1.5 kHz) single-channel currents were sampled at 5 kHz on a PDP-11/73 computer. All openings above a threshold equal to 50% of the single-channel amplitude were included in the computation for total current. To enable rapid, on-line computation of NP_o, the single-channel current data was converted to a binary mask (currents under the threshold were assigned a value of 0 while those above the threshold were assigned a value of 1). The binary mask was then multiplied with the original single-channel current data and integrated to yield a "total" current, I, which included only openings above the threshold. This current, I, was divided by the single-channel current, i, to yield NP_o.

Preparation of G Protein Subunits

$G_{\beta\gamma}$ suspended in Lubrol and provided by Dr. P. J. Casey (University of Texas, Southwestern Medical School, Dallas, TX) was prepared as described previously (Roof et al., 1985) with modifications as follows. After the final heptylamine Sepharose column, the $G_{\beta\gamma}$ and G_{α} were resolved on a fast protein liquid chromatogaphy (FPLC) Mono-Q column (Pharmacia Inc., Piscataway, NJ) in the presence of $AlCl_3$, $MgCl_2$, and NaF. The preparation was injected onto a 0.5 × 5.0 cm Mono-Q column equilibrated in 50 mM Tris HCl (pH 8.0), 1 mM EDTA, 1 mM dithiothreitol (DTT), 1.0% sodium cholate, 20 μM $AlCl_3$, 6 mM $MgCl_2$, and 10 mM NaF. The column was eluted with a 30 ml gradient of 0–250 mM NaCl in the same buffer. $G_{\beta\gamma}$ eluted at ~170 mM. The Mono-Q pool of $G_{\beta\gamma}$ was then diluted 10-fold into 20 mM HEPES, pH 7.6, 25 mM NaCl, 1.0% cholate (buffer A) and chromatographed on a hydroxylapatite column using a linear gradient of buffer A to buffer A + 80 mM potassium phosphate. $G_{\beta\gamma}$ eluted at 40 mM phosphate and was concentrated using a Centricon 30 to 2.5 mg/ml. The sample was then gel-filtered (Sephadex G-50) into 50 mM HEPES, 1 mM DTT, 1 mM EDTA, and 0.1% Lubrol.

Porcine cardiac $G_{\beta\gamma}$ subunits supplied by Dr. M. Schimerlik (Oregon State University, Corvallis, OR) were prepared by a modification of the procedure for the preparation of G_i subunits described by Katada et al. (1986). G_i (50 μg) was incubated for 2 h at 32°C in a final volume of 250 μl containing 10 μM GTPγS, 10 mM Na-HEPES (pH 7.4), 0.1 M NaCl, 5 mM $MgCl_2$, 1 mM EGTA, 1 mM DTT, 0.1 mM phenyl-methyl-sulfonyl-fluoride and 0.1% vol/vol Lubrol PX. The activated protein was diluted 20-fold with cold buffer A containing 0.1 M NaCl and 0.25% wt/vol cholate and applied to a 10-ml column of heptyl agarose. The column was eluted with 10 ml of buffer A plus 0.1 M NaCl and 0.25% wt/vol cholate followed by a 60-ml gradient from 0.25% wt/vol cholate, 0.25 M NaCl to 1.2% wt/vol cholate, 0.025 M NaCl in buffer A. $G_{i\alpha}$ did not bind to the column and was eluted before the application of the gradient while $G_{i\beta\gamma}$ was eluted at ≈0.8% wt/vol cholate.

Transducin $G_{\beta\gamma}$ subunits ($G_{Tr\beta\gamma}$) were provided by Dr. G. L. Johnson (National Jewish Center, Denver, CO) and were prepared as described previously (Wessling-Resnick and Johnson, 1987; Kelleher and Johnson, 1988) with the following modifications. The G_{α} and $G_{\beta\gamma}$ were resolved on a FPLC column, and the samples were concentrated overnight under vacuum at 0–4°C in a Pro Di Con (Biomolecular Dynamics, Beaverton, OR) concentrator. G_{α} and $G_{\beta\gamma}$ subunits, which were applied to guinea pig atrial cells (Figs. 6 and 7 A), were provided by Drs. I. Kobayashi, T. Katada, and M. Ui (Katada et al., 1986); monoclonal antibody 4A against transducin α ($G_{Tr\alpha}$) was provided by Dr. H. E. Hamm (University of Illinois College of Medicine, Chicago, IL), and was prepared as described previously (Deretic and Hamm, 1987; Hamm et al., 1987).

Results

Effects of Detergents on $G_{\beta\gamma}$ Activation of $i_{K.ACh}$

The inwardly rectifying cardiac atrial channel, $i_{K.ACh}$, is a 40–50-pS (mean open time, ~1 ms), K^+-selective channel that is activated by ACh, somatostatin, and adenosine, via PTX-sensitive G proteins. Both subunits of PTX-sensitive G proteins can activate this channel when applied to the intracellular side of inside-out patches (the bath). The relatively hydrophilic G_α subunits will remain in suspension in a physiological buffer while the hydrophobic $G_{\beta\gamma}$ subunits of most G proteins must be suspended in a detergent such as CHAPS or Lubrol PX to prevent aggregation. Much of the controversy on activation by $G_{\beta\gamma}$ has centered around activation by the detergent alone. Low concentrations (≤ 500 μM) of the zwitterionic detergent, CHAPS, had no effect on the channel while higher concentrations caused patch breakdown. Therefore, in initial experiments, $G_{\beta\gamma}$ was suspended in 180–200 μM CHAPS. Although ~80% of the subunits were aggregated at these low concentrations of detergent (Logothetis et al., 1988), the subunit still activated the channel. To verify that this activation was not caused by the unique combination of CHAPS and $G_{\beta\gamma}$, $G_{\beta\gamma}$ suspended in Lubrol PX (a nonionic detergent) was applied to inside-out patches. In the cell-attached configuration, with ACh in the pipette, $i_{K.ACh}$ was active (Fig. 3 A, left). Under the conditions of these experiments (symmetrical K^+ on both sides of the patch), the reversal potential for K^+ was 0 mV (with a holding potential of -80 mV, the K^+ current is inward). When an inside-out patch was formed, $i_{K.ACh}$ activity rapidly declined due to washout of GTP. Exposure of the patch to 300 nM Lubrol PX for 5 min had no effect on channel activity. Channel activity increased 2 min after exposing the patch to 10 nM $G_{\beta\gamma}$ in Lubrol. $G_{\beta\gamma}$ maximally activated the patch since GTP added after $G_{\beta\gamma}$ failed to increase the activity any further (12 of 18 patches).

While low concentrations (300 nM) of Lubrol PX had no effect on the channel (Fig. 3 A, higher concentrations (16 μM) blocked the preactivated channel (Fig. 3 B). Addition of GTPγS, a nonhydrolyzable GTP analogue, to the bath irreversibly activates G proteins and hence $i_{K.ACh}$, in inside-out patches. In control experiments, GTPγS-induced activation was not reversible over a period of >30 min when the patch was perfused with saline alone. When 16 μM (0.001%) Lubrol PX alone was added to the intracellular side of such an irreversibly activated inside-out patch from neonatal rat atria, channel activity rapidly declined. Lubrol PX-dependent inhibition of GTPγS-induced activation was observed in four of our patches tested with 16 μM Lubrol PX. In summary, only lower concentrations of detergents (< 200 μM for CHAPS, or < 1 μM for Lubrol) could be used to suspend the $G_{\beta\gamma}$ subunits. $G_{\beta\gamma}$ suspended in either detergent activated the channel while denatured (boiled) $G_{\beta\gamma}$ in these detergents failed to activate the channel (Logothetis et al., 1987; Nanavati, C., and D. E. Clapham, unpublished observations). We conclude that channel activation by $G_{\beta\gamma}$ was caused by the protein itself.

Activation by Cardiac $G_{\beta\gamma}$

G protein subunits used for reconstitution studies have been purified from several sources including erythrocytes, placenta, and brain. To test for tissue specificity, we applied $G_{\beta\gamma}$ purified from porcine heart to inside-out patches (Fig. 4). The frequency of channel openings in this patch was low under control conditions since the endogenous G proteins in this patch were in an inactive state (neither GTP nor GTPγS were present). Approximately 1 min after 10 mM porcine cardiac $G_{\beta\gamma}$ was added to the bath, $i_{K.ACh}$ was maximally activated. Such activation occurred in 6 of 10 experiments (neonatal rat and embryonic chick atrial cells). The characteristics of the channels (open time, single-channel conductance) activated by cardiac $G_{\beta\gamma}$ were indistinguishable from the characteristics of channels activated with $G_{\beta\gamma}$ from other tissues.

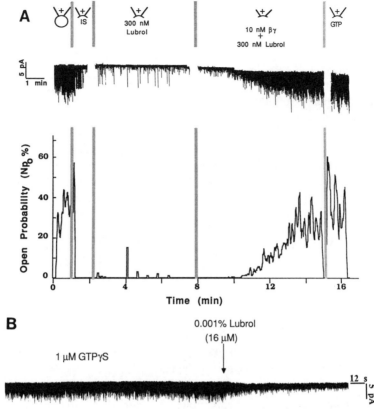

Figure 3. Activation by $G_{\beta\gamma}$ was independent of the detergent used to suspend the subunits. (*A*) In the cell-attached configuration (*left*), with ACh in the pipette, $i_{K.ACh}$ activity was high as shown both by the current trace (*top*), and by the channel activity (NP_o; *bottom*). When an inside-out patch was formed, $i_{K.ACh}$ activity decreased due to the washout of GTP, while $i_{K.ATP}$ activity (the infrequent 6-pA openings caused by activation of the ATP depletion–dependent channel) increased due to the washout of ATP. Exposure to 300 nM Lubrol for up to 5 min had no effect (transient increases in NP_o are caused by openings of $i_{K.ATP}$). 10 nM $G_{\beta\gamma}$ in Lubrol (provided by Dr. P. J. Casey) activated the channel maximally because the addition of GTP at the end of the experiment failed to further activate the channel. (*B*) Inhibition by high concentrations of Lubrol PX. This inside-out patch was irreversibly activated by 1 μM GTPγS. The addition of 16 μM Lubrol to the bath rapidly blocked GTPγS-induced channel activity (Fig. 3 *B* provided by Dr. D. Kim).

Effects of Transducin βγ Subunits

Two well-studied G protein–linked receptor systems are the β-adrenergic receptor-G_s-adenylyl cyclase system, and the rhodopsin-G_{Tr}-cGMP phosphodiesterase system. In both these systems, receptor binding causes effector stimulation via the activated G_α subunit. Although not supported by our results, it has been suggested that the $G_{\beta\gamma}$ subunit plays an inhibitory or a modulatory role in $i_{K.ACh}$ activation. If $G_{\beta\gamma}$ merely played an inhibitory role, exogenously added $G_{\beta\gamma}$ would bind endogenous G_α, thereby reducing channel activity. To avoid possible complications with detergents, the more hydrophilic, $G_{\beta\gamma}$ subunits from transducin ($G_{Tr\beta\gamma}$) were used for this experiment.

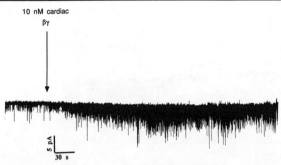

Figure 4. Activation of $i_{K.ACh}$ by cardiac $G_{\beta\gamma}$ (provided by Dr. M. Schimerlik). The basal single-channel $i_{K.ACh}$ activity in this inside-out patch was low in the absence of GTP or an analogue. When 10 nM cardiac $G_{\beta\gamma}$ (suspended in CHAPS) was added to the intracellular surface of the patch (i.e., to the recording chamber), channel activity increased. The infrequent 6-pA channel openings in this patch were caused by the ATP depletion–activated K$^+$ channel, $i_{K.ATP}$ (figure provided by Dr. D. Kim).

When GTP was added to an inside-out patch from a neonatal rat atrial cell (Fig. 5; ACh in the pipette), channel activity rapidly increased. Subsequent addition of 10 nM $G_{Tr\beta\gamma}$ had no significant effect on channel activity. In 16 experiments, $G_{Tr\beta\gamma}$ from three different batches failed to inhibit $i_{K.ACh}$. In 2 of 11 patches, $G_{Tr\beta\gamma}$ (3–10 nM) activated $i_{K.ACh}$.

G_α Activation of $i_{K.ACh}$

In excised patches, both the G_α and the $G_{\beta\gamma}$ subunit activated $i_{K.ACh}$. However, it was uncertain which (or if both) subunit(s) was involved in the physiological coupling between the muscarinic receptor and the channel. This issue was addressed by Yatani et al. (1989) who reported that a monoclonal antibody (MAb4A) made against the G_α subunit of transducin blocked the coupling between the muscarinic receptor and $i_{K.ACh}$ in adult guinea pig atrial cells. In the experiments reported, the high $i_{K.ACh}$ activity observed in inside-out patches exposed to GTP (in the presence of agonist) was rapidly abolished when MAb4A was introduced to the bath. This inhibition of the endogenous G protein–effector coupling required that the G protein be in its active GTP-bound state. Furthermore, the inhibition was irreversible since the channel could not be activated by GTP or GTPγS even after the antibody had been removed from the chamber. On the basis of these experiments, they concluded that the G_α subunit was

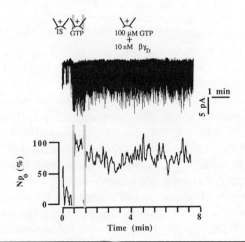

Figure 5. Transducin $G_{\beta\gamma}$ (10 nM; provided by Dr. G. L. Johnson) failed to inhibit $i_{K.ACh}$ in a patch previously activated by 100 μM GTP (1 μM ACh in the pipette).

the signal transducer in vivo (Yatani et al., 1989). We attempted to reproduce these experiments in neonatal rat and adult guinea pig atrial cells. Fig. 6. shows the single-channel currents recorded from two different inside-out patches from adult guinea pig atrial cells. In both instances, 0.1 μM ACh was included in the pipette and 10 μM GTP was added to the bath. The top trace (Fig. 6 A) is from a patch in which exposure to MAb4A for 5 min failed to block GTP activation. Activity gradually subsided as GTP was washed out but could be rapidly restored by adding 10 μM GTPγS. In 6 of 12 inside-out patches, the antibody had similar effects, and clearly failed to block $i_{K.ACh}$. In the remaining six patches addition of MAb4A slowly reduced $i_{K.ACh}$ activity, but upon addition of GTPγS the channels were reactivated (Fig. 6 B). Thus the reduction in channel activity observed in these latter six experiments was reversible, and was probably not a specific blocking effect of the antibody.

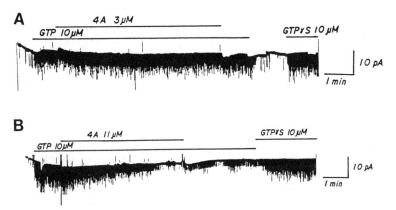

Figure 6. MAb4A failed to block $i_{K.ACh}$. In two different experiments, with 0.1 μM ACh in the pipette, $i_{K.ACh}$ in inside-out patches was activated by the addition of 10 μM GTP to the bath. (A) Addition of 3 μM MAb4A (provided by Dr. H. E. Hamm) to the bath failed to inhibit channel activity (up to 5 min). As expected, GTP removal reduced activity, which could be increased upon addition of GTPγS. (B) 11 μM MAb4A slightly reduced GTP-induced activity. When the antibody was removed, thereby exposing the patch to GTP alone, activity increased. Maximal activation could be achieved by exposing the patch to 10 μM GTPγS.

A Comparison of the Activation of $i_{K.ACh}$ by G_α and $G_{\beta\gamma}$

Codina et al. (1987) reported that 50 pM of $G_{\alpha i3}$ activated the patch maximally. These results contrasted those of Kurachi et al. (1989). The experiments shown in Fig. 7 A were designed to compare $i_{K.ACh}$ activation by G_α and $G_{\beta\gamma}$ subunits. The three traces show the currents from three experiments in which the channels in inside-out patches from adult guinea pig atrial cells were activated by the addition of GTP to the bath (the pipette contained 1 μM ACh). The GTP was then washed out and 10 nM of either $G_{\alpha_{39}}$, $G_{\alpha_{40}}$, or $G_{\alpha_{41}}$ was added to the chamber. In the experiments shown, $G_{\alpha_{39}}$ had little or no effect, while both $G_{\alpha_{40}}$ and $G_{\alpha_{41}}$ partially activated $i_{K.ACh}$. 10 nM $G_{\beta\gamma}$ was added once channel activation by the G_α subunits had reached a steady-state level. Steady-state channel activity with the $G_{\beta\gamma}$ subunits was equivalent to intrinsic activation by GTP (ACh in the pipette). This additional activation by $G_{\beta\gamma}$ (10 nM) could not be due to G_α contamination since activation of the channel by 10 nM G_α subunits added alone had

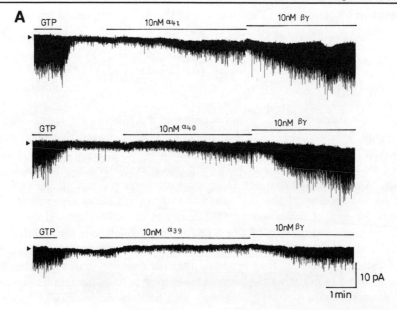

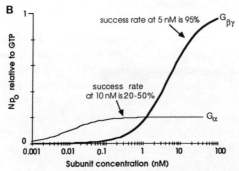

Figure 7. Relative ability of activated G_α vs. $G_{\beta\gamma}$ to activate $i_{K.ACh}$. (*A*) Application of GTP to the bath caused maximal activation of $i_{K.ACh}$ in inside-out patches (1 μM ACh in the pipette). After the removal of GTP, the addition of 10 nM of the indicated G_α subunit (provided by Drs. Katada, Ui, and Kobayashi) caused partial ($G_{\alpha40}$ and $G_{\alpha41}$) or no ($G_{\alpha39}$) activation of the channel. Subsequent addition of 10 nM $G_{\beta\gamma}$ activated the channel maximally. Experiments shown were performed on atria from adult guinea pigs. (*B*) A comparison of the dose-response curves for the G_α and $G_{\beta\gamma}$ subunits. The sigmoidal curves are best fits to data. The activation by the G_α subunits occurs at lower concentration, but less reproducibly than the activation by $G_{\beta\gamma}$ subunits.

already reached steady-state levels. A comparison of the dose-response curves for the two G protein subunits (Fig. 7 *B*) shows that picomolar concentrations of G_α subunits and nanomolar concentrations of $G_{\beta\gamma}$ subunits are necessary for channel activation. Since the relative ability of subunits to access the membrane is unknown, concentrations of subunits that activate $i_{K.ACh}$ do not settle the issue of which subunit activates the channel in vivo. The degree of activation by nanomolar concentrations of G_α varies widely (see Logothetis et al., 1988; Kurachi et al., 1989; and Fig. 7 *A*) while similar concentrations of $G_{\beta\gamma}$ always activate the channel maximally. Finally, the frequency of activation of the channel by $G_{\beta\gamma}$ is more reproducible (>95% success) than activation by G_α (20–50% success).

Discussion

The report by Logothetis et al. (1987) was one of the first reports demonstrating a direct role for $G_{\beta\gamma}$ in signal transduction. Since then, $G_{\beta\gamma}$ has been described as an important messenger in other systems: (a) Katada et al. (1987) showed that the Ca^{2+}-calmodulin activation of rat brain adenylyl cyclase could be inhibited by 2–10 nM porcine brain $G_{i\beta\gamma}$ or $G_{o\beta\gamma}$. They postulated that this inhibition was due to the formation of a calmodulin-$G_{\beta\gamma}$ complex that was incapable of activating cyclase. (b) Jelsema and Axelrod (1987) and Jelsema et al. (1989) showed that $G_{Tr\beta\gamma}$ activated phospholipase $A_2(PLA_2)$ in rod outer segments. Since the $G_{Tr\beta\gamma}$ activation of PLA_2 occurred even in the presence of GTPγS which prevents reassociation of G protein subunits, they concluded that the $G_{Tr\beta\gamma}$ activation was a direct effect, and was not caused by the binding of the $G_{\beta\gamma}$ subunits with an inhibitory G_α subunit, thereby causing a removal of inhibition. (c) Whiteway et al. (1989) showed that in yeast, genetic mutants lacking the STE4 or STE18 genes (which code for putative yeast G protein β and γ subunits, respectively), were unable to respond to pheromone. On the basis of these experiments, they proposed that the putative yeast $G_{\beta\gamma}$ subunits were directly involved in initiating the pheromone response. Also, in mutants in which the SCG1 gene (putative yeast G_α subunit) had been modified or was lacking, the pheromone pathway was constitutively activated, suggesting that the excess free $G_{\beta\gamma}$ stimulated the pathway.

In the cardiac muscarine-gated $i_{K.ACh}$ system, both the α and the $\beta\gamma$ subunits of G proteins are capable of activating an effector. The activation by $G_{\beta\gamma}$ is attributable neither to the detergent nor to contamination by activated G_α subunits and may be via a mechanism that is distinct from G_α activation of $i_{K.ACh}$. The reasons for the variability in the activation of the channel by G_α (activation occurs in 20–50% of the inside-out patches containing the channel) is unknown, but may be due to partially denatured protein or to the absence of intermediates necessary for channel activation. It is worth noting that the various G_α subunit subtypes from different sources (from Dr. Neer, see Logothetis et al., 1987, 1988; Dr. Birnbaumer, see Logothetis et al., 1988; and Drs. Katada and Ui, see Kurachi et al., 1989, and this paper) vary in their abilities to activate the channel.

Brown, Birnbaumer, and colleagues have reported that CHAPS activates the channel while $G_{Tr\beta\gamma}$, brain $G_{\beta\gamma}$ in Lubrol, and an antibody against transducin all inhibit the channel. In the experiments presented here, $G_{Tr\beta\gamma}$ (which had never been exposed to detergent) occasionally activated and always failed to inhibit $i_{K.ACh}$ once the channel had been activated by intracellular application of GTP in the presence of ACh. Also, $G_{\beta\gamma}$ suspended in low concentrations of Lubrol always activated and never inhibited the channel. Finally, MAb4A did not inhibit $i_{K.ACh}$ activation by GTP in either adult guinea pig or neonatal rat atria. Thus, it is still unclear whether G_α, $G_{\beta\gamma}$, or both are the physiological activators of the channel.

At least 17 G_α, 4 G_β, and 3 G_γ subunits have been identified by cloning techniques. Therefore, 12 different $G_{\beta\gamma}$ subunits are possible on the basis of sequence alone. Gautam et al. (1989) described a brain G_γ clone with a *ras*-like consensus sequence for palmitoylation. Like the G_α subunits (Buss et al., 1987; also see references in Lowy and Willumsen, 1989), the G protein β and/or γ subunits may be posttranslationally modified, thus producing more potential functional diversity. Given the diversity of subtypes and the extensive homology within a specific subtype, it may not be possible to assign specific function to G protein subunits by reconstitution experiments alone. In this respect, molecular biology may provide the tools to modify native subunits in ways that allow specific assignment of their roles.

Acknowledgments

We are grateful to Drs. P. J. Casey, H. E. Hamm, G. L. Johnson, T. Katada, I. Kobayashi, M. Schimerlik, and M. Ui for their gifts of G protein subunits and antibodies for experiments presented in this paper; and to Dr. D. Kim for the experiments shown in Figs. 3 B and 4. We thank Dr. Eva Neer for helpful advice and comments, Rose Bengal for culturing the neonatal rat atrial cells, and Chris Bliton for writing the on-line NP_o analysis program.

This work was supported by National Institutes of Health grant HL-41303 and an American Heart Association Established Investigator Award to D. E. Clapham and grants from the Ministry for Education, Sciences, and Culture of Japan to Y. Kurachi.

References

Breitwieser, G. E., and G. Szabo. 1985. Uncoupling of cardiac muscarinic and β-adrenergic receptors from ion channels by a guanine nucleotide analogue. *Nature.* 317:538–540.

Buss, J. E., S. M. Mumby, P. J. Casey, A. G. Gilman, B. M. Sefton. 1987. Myristoylated α subunits of guanine nucleotide-binding regulatory proteins. *Proceedings of the National Academy of Sciences.* 84:7493–7497.

Cerbai, E., U. Klockner, and G. Isenberg. 1988. The α-subunit of the GTP-binding protein activates muscarinic potassium channels of the atrium. *Science.* 240:1782–1783.

Codina, J., A. Yatani, D. Grenet, A. M. Brown, and L. Birnbaumer. 1987. The α-subunit of the GTP binding protein G_k opens atrial potassium channels. *Science.* 236:442–445.

Deretic, D., and H. E. Hamm. 1987. Topographic analysis of antigenic determinants recognized by monoclonal antibodies to the photoreceptor guanyl nucleotide-binding protein, transducin. *Journal of Biological Chemistry.* 262:10839–10847.

Gautam, N., M. Baetscher, R. Aebersold, and M. I. Simon. 1989. A G protein gamma subunit shares homology with ras proteins. *Science.* 244:971–974.

Hamill, O. P., A. Marty, E. Neher, B. Sakmann, and F. J. Sigworth. 1981. Improved patch-clamp techniques for high-resolution current recording from cells and cell-free membrane patches. *Pflügers Archiv.* 391:85–100.

Hamm, H. E., D. Deretic, K. P. Hofmann, A. Schleicher, and B. Kohl. 1987. Mechanism of action of monoclonal antibodies that block the light activation of the guanyl nucleotide-binding protein, transducin. *Journal of Biological Chemistry.* 262:10831–10838.

Hartzell, H. C. 1988. Regulation of cardiac ion channels by catecholamines, acetylcholine and second messenger systems. *Progress in Biophysics and Molecular Biology.* 52:165–247.

Jelsema, C. L., and J. Axelrod. 1987. Stimulation of phospholipase A2 activity in bovine rod outer segments by the beta gamma subunits of transducin and its inhibition by the alpha subunit. *Proceedings of the National Academy of Sciences.* 84:3623–3627.

Jelsema, C. L., R. M. Burch, S. Jaken, A. D. Ma, and J. Axelrod. 1989. Modulation of phospholipase A_2 activity in rod outer segments of bovine retina by G protein subunits, guanine nucleotides, protein kinases and calpactin. *Neurology and Neurobiology.* 49:25–41.

Katada, T., K. Kusakabe, M. Oinuma, and M. Ui. 1987. A novel mechanism for the inhibition of adenylate cyclase via inhibitory GTP-binding proteins. Calmodulin-dependent inhibition of the cyclase catalyst by the $\beta\gamma$ subunits of GTP binding proteins. *Journal of Biological Chemistry.* 262:11897–11900.

Katada, T., M. Oinuma, and M. Ui. 1986. Two guanine nucleotide-binding proteins in rat brain serving as the specific substrate of islet-activating protein, pertussis toxin. Interaction of the α-subunits with $\beta\gamma$-subunits in development of their biological activities. *Journal of Biological Chemistry.* 261:8182–8191.

Kelleher, D. J., and G. L. Johnson. 1988. Transducin inhibition of light-dependent rhodopsin phosphorylation evidence for $\beta\gamma$ subunit interaction with rhodopsin. *Molecular Pharmacology.* 34:452–460.

Kurachi, Y. 1989. Regulation of G protein-gated K^+ channels. *News in Physiological Sciences.* 4:158–161.

Kurachi, Y., H. Ito, T. Sugimoto, T. Katada, and M. Ui. 1989. Activation of atrial muscarinic K^+ channels by low concentrations of $\beta\gamma$ subunits of rat brain G protein. *Pflügers Archiv.* 413:325–327.

Kurachi, Y., T. Nakajima, and T. Sugimoto. 1986a. Acetylcholine activation of K channels in cell-free membranes of atrial cells. *American Journal of Physiology.* 251:H681–684.

Kurachi, Y., T. Nakajima, and T. Sugimoto. 1986b. On the mechanism of activation of muscarinic K^+ channels by adenosine in isolated atrial cells: involvement of GTP-binding proteins. *Pflügers Archiv.* 407:264–274.

Lewis, D. L., and D. E. Clapham. 1989. Somatostatin activates an inwardly rectifying K^+ channel in neonatal rat atrial cells. *Pflügers Archiv.* 414:492–494.

Logothetis, D. E., D. Kim, J. K. Northup, E. J. Neer, and D. E. Clapham. 1988. Specificity of action of guanine nucleotide-binding regulatory protein subunits on the cardiac muscarinic K^+ channel. *Proceedings of the National Academy of Sciences.* 85:5814–5818.

Logothetis, D. E., Y. Kurachi, J. Galper, E. J. Neer, and D. E. Clapham. 1987. The $\beta\gamma$ subunits of GTP-binding proteins activate the muscarinic K^+ channel in heart. *Nature.* 325:321–326.

Lowy, D. R., and B. M. Willumsen. 1989. New clue to ras lipid glue. *Nature.* 341:384–385.

Pfaffinger, P. J., J. M. Martin, D. D. Hunter, N. M. Nathanson, and B. Hille. 1985. GTP binding proteins couple cardiac muscarinic receptors to a K channel. *Nature.* 317:536–538.

Pusch, M., and E. Neher. 1988. Rates of diffusional exchange between small cells and a measuring patch pipette. *Pflügers Archiv.* 411:204–211.

Roof, D. J., M. L. Appleton, and P. C. Sternweis. 1985. Relationships within the family of GTP-binding proteins isolated from bovine central nervous system. *Journal of Biological Chemistry.* 260:16242–16249.

Wessling-Resnick, M., and G. L. Johnson. 1987. Kinetic and hydrodynamic properties of transducin: comparison of physical and structural parameters for GTP-binding regulatory proteins. *Biochemistry.* 26:4316–4323.

Whiteway, M., L. Hougan, D. Dignard, D. Y. Thomas, L. Bell, G. C. Saari, F. J. Grant, P. O'Hara, and V. L. MacKay. 1989. The STE4 and STE18 genes of yeast encode potential β and γ subunits of the mating factor receptor-coupled G protein. *Cell.* 56:467–477.

Yatani, A., H. Hamm, J. Codina, M. R. Mazzoni, L. Birnbaumer, and A. M. Brown. 1989. A monoclonal antibody to the α subunit of G_k blocks muscarinic activation of atrial K^+ channels. *Science.* 241:828–831.

Chapter 4

Modulation of M Current in Frog Sympathetic Ganglion Cells

Martha M. Bosma, Laurent Bernheim, Mark D. Leibowitz, Paul J. Pfaffinger, and Bertil Hille

Department of Physiology and Biophysics, University of Washington School of Medicine, Seattle, Washington 98195

Introduction

At many synapses fast excitatory transmission is mediated by agonist-gated channels and modulated by slower actions via G protein–coupled receptors. Synaptic transmission in frog lumbar sympathetic ganglia is a well-studied example. There, according to classical physiology known for several generations, cholinergic preganglionic fibers excite postganglionic cells via large, fast excitatory postsynaptic potentials (EPSPs) generated by nicotinic acetylcholine receptors (nAChR). As the EPSP elicited by one input fiber is typically suprathreshold for exciting a postganglionic cell, transmission is virtually guaranteed. In addition, however, the preganglionic fibers and perhaps some sensory fibers modulate the excitability of postganglionic cells. A synaptic sensitization accompanied by a small, slow EPSP is mediated by muscarinic acetylcholine receptors (mAChR) and by receptors for luteinizing hormone-releasing hormone (LHRH = GnRH) and substance P (SP) (Kuba and Koketsu, 1976; Jan and Jan, 1982). The end result is that the postganglionic cell may fire a burst of spikes rather than a single spike in response to subsequent fast EPSPs.

A primary modulatory action of the three transmitters acetylcholine (ACh), LHRH, and SP is to close a noninactivating, voltage-gated K channel, the M-type K channel (Adams and Brown, 1980; Brown and Adams, 1980; Jan and Jan, 1982; Adams et al., 1983; reviewed by Brown, 1988). Shutting M-type channels facilitates fast excitatory transmission in two ways: it increases the input resistance of the postsynaptic cell and also depolarizes the membrane somewhat (slow EPSP). Although other currents are modulated by these neurotransmitters (Kuffler and Sejnowski, 1983; Brown and Selyanko, 1985), the suppression of M current (I_M) has received by far the most attention.

Like the laboratories of P. R. Adams and D. A. Brown, ours has been asking which intracellular signaling mechanisms couple transmitter action to modulation of I_M. This chapter reviews our published (Pfaffinger, 1988; Pfaffinger et al., 1988; Bosma and Hille, 1989) and unpublished work on frog lumbar sympathetic ganglion cells.

We developed a preparation using the whole-cell clamp configuration of the patch-clamp technique to study large (B) postganglionic cells acutely isolated by a combination of enzymatic and mechanical treatments of desheathed ganglia (Pfaffinger, 1988). The whole-cell method has an advantage over the microelectrode methods used before in that it allows ready delivery of molecules into the cytoplasm. I_M is recorded using the classical protocol of Brown and Adams (1980), holding the membrane potential at -35 mV where all outward K currents except I_M inactivate. To verify that the standing outward current at this potential is primarily I_M, the voltage-dependent gating on this conductance is checked periodically with hyperpolarizing pulses. Many of our records show this 400–800-ms step to -60 mV applied every 5 s. It induces a characteristic, slow closing of M-type channels at -60 mV and a reopening at -35 mV.

Agonists Have Three Clear Electrophysiological Effects

I_M of ganglion cells changes in several characteristic ways in response to modulatory transmitters. We distinguish three processes: suppression, desensitization, and overrecovery. Fig. 1 A shows two of these effects during an application of 1 µM teleost LHRH (t-LHRH). Before the peptide is applied, there is a standing outward current of 300 pA, which is primarily I_M as judged by the large closing and reopening transients elicited by the hyperpolarizing pulses (shown expanded in time in Fig. 1, *inset 1*). When t-LHRH is applied (bar above), I_M becomes suppressed as can be recognized by the decrease of standing current and the reduction of the gating transients (Fig. 1, *insets 2* and *3*). The reduction persists during the 5 min that t-LHRH is present and is more than fully reversed when the peptide is washed away (Fig. 1, *inset 4*). We call this "overrecovery." It nearly always occurs for a couple of minutes after application of LHRH, SP, or muscarine. The cycle of suppression and recovery can be repeated with subsequent applications of 1μM t-LHRH (not shown). Fig. 1 *B* shows the effects of a high concentration of a more potent agonist, chicken II LHRH (cII-LHRH). It suppresses I_M too, but after a minute, I_M begins to return even though cII-LHRH remains in the bath. This is densitization. It produces a persistent loss of sensitivity to subsequent applications of the agonist.

We would like to determine which signaling pathways give rise to suppression, desensitization, and overrecovery. The possibilities can be addressed in the context of a menu (Fig. 2, *left*) showing some components of cellular signaling systems. At the top

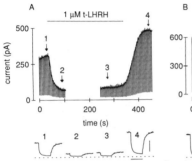 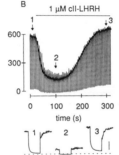

Figure 1. The three actions of agonists on I_M. (*Upper traces*) Total membrane current recorded for 5–7 min. Agonist is applied in the bath for the period marked with a bar. Membrane potential is held at −35 mV, except every 5 s it is stepped to −60 mV for 600 ms. (*Lower traces*) Total membrane current on a faster time scale during the hyperpolarizing test steps indicated by arrows above. (Bars = 200 pA, 600 ms). (*A*) 1 μM t-LHRH suppresses I_M and there is overrecovery when it is washed off. (*B*) 1 μM cII-LHRH suppresses I_M, desensitizes the response, and leads to overrecovery while agonist is still present. The recordings are done with MgATP, GTP, leupeptin, EGTA, buffer and K salts in the pipette at room temperature. (From Bosma and Hille, 1989.)

are a variety of receptors; muscarinic, SP, and LHRH receptors have been circled for this study. At the bottom are possible targets; the M-type K channel is the focus here but in addition two classes of Ca channel have been circled to indicate that they are also modulated by SP and LHRH in frog sympathetic ganglion cells (Bley and Tsien, 1988; Jones and Marks, 1989). The task is to define how the signal passes from receptors to targets. The menu shows a list of GTP-binding proteins (G proteins), several enzymes that can be activated by G proteins, second messenger molecules that might be produced, and protein kinases that respond to the second messengers.

Early work argued against cyclic AMP and cyclic GMP as mediators of the signal (Adams et al., 1982), a conclusion that we have confirmed (unpublished observations). I_M and its coupling to agonists are not altered by injections of cyclic nucleotides from micropipettes, by putting cyclic nucleotides in whole-cell pipettes or permeant and poorly hydrolyzable derivatives in the bath, or by adding forskolin to the bath. Therefore in this chapter we focus on other pathways.

A Pertussis Toxin–insensitive G Protein Is Involved

Our standard recording conditions include 100 μM GTP in the "intracellular" pipette solution to preserve signaling pathways that require G proteins. The role of G proteins in suppression of I_M can be studied by including GTP analogues or G protein activators in the pipette as well. Three agents, GTPγS, F$^-$, and AlF$_4^-$, can suppress I_M fully without the addition of modulatory transmitter to the bath. The time constant of the loss of I_M is 120 s with 50 μM GTPγS, 50 s with 10 mM NaF, and 12 s with 10 mM NaF + 100 μM AlCl$_3$ (Pfaffinger, 1988). Presumably each of these agents is activating a G protein that can suppress I_M. GTPγS would act the slowest because it must wait for a spontaneous dissociation of GDP from the G protein, whereas AlF$_4^-$ can act directly on the GDP-bound form of the G protein (Sternweis and Gilman, 1982). Nevertheless, the spontaneous action of GTPγS is relatively quick, implying either that the relevant class of G proteins cycles GDP fairly frequently (minutes) or that the assumption that

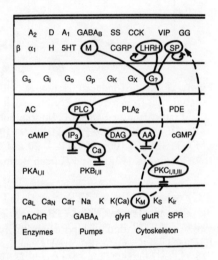

Figure 2. A menu approach to analyzing second-messenger pathways. (A) The starting and end points are marked. (B) The results of this study are filled in.

there are no agonist molecules in the Ringer's solution is wrong. For comparison, in other electrophysiological experiments on intact cells or membrane patches the G protein that couples to phospholipase C (PLC) in lacrimal gland and that which couples to K_{ACh} channels in cardiac atrium can be activated by GTPγS about this fast, whereas transducin in photoreceptors seems not to be activated at all (Evans and Marty, 1986; Kurachi et al., 1986; Rispoli, G., and P. B. Detwiler, unpublished observations).

The foregoing experiments show that a G protein couples to suppression of I_M, but they do not show if the receptor-mediated response proceeds this way. Other experiments with GTP analogues do (Fig. 3). In Fig. 3 A the agonist-induced suppression of I_M is reversible and overrecovers when the intracellular pipette contains 100 μM GTP, but it becomes partly irreversible when the pipette contains 100 μM GPP(NH)P (Fig.

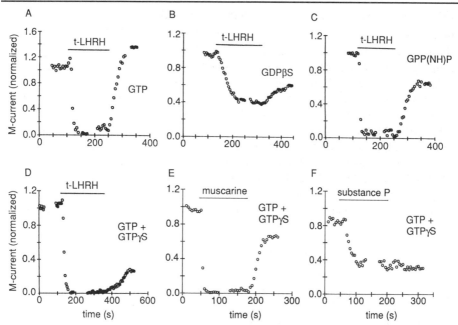

Figure 3. Agonist-induced responses with different GTP analogues in the pipette. Each experiment is from a different cell tested at least 8 min after breakthrough with 1 μM t-LHRH, 20 μM muscarine, or 50 nM SP. Internal solutions: (*A*) 100 μM GTP; (*B*) 100 μM GDPβS + 20 μM GTP; (*C*) 100 μM GPP(NH)P; (*D*) 100 μM GTP + 1 μM GTPγS; (*E*) 100 μM GTP + 4 μM GTPγS; (*F*) 100 μM GTP + 20 μM GTPγS. Leak currents have been subtracted. (*A–D* from Pfaffinger, 1988; *E* and *F*, from Bosma, M. M., unpublished).

3 *C*) or a mixture of GTPγS with GTP (Fig. 3, *D–F*).[1] Presumably this means that agonists accelerate GDP-GTP exchange on a G protein so that the nucleotide-binding site becomes loaded with poorly hydrolyzable GTPγS or GPP(NH)P and the G protein becomes persistently activated. Consistent with this hypothesis, the response to t-LHRH is partially blocked when the pipette contains 100 μM GDPβS (Fig. 3 *B*). Thus the receptor-mediated suppression of I_M requires a G protein. Since such tests with nucleotide analogues block or make suppression irreversible, we do not have a protocol that allows us to ask if overrecovery or desensitization requires G proteins.

We have not identified the G protein involved. As is often true of muscarinic responses in other systems (Nathanson, 1987), this one is not pertussis toxin inhibited (Pfaffinger, 1988).[2] Possible stimulation by cholera toxin has not been tested. Some features of Fig. 3 may further typify this G protein (but alternatively may result from actions of other intracellular enzymes in these cells, e.g., Otera et al., 1988). These

---

[1] Mixing GTPγS with higher concentrations of GTP is a useful method to greatly reduce the spontaneous activation of G proteins without preventing the rapid accumulation of GTPγS-activated G proteins during the fast cycle induced by agonist (Breitwieser and Szabo, 1988).

[2] The response was not prevented by incubating cells with pertussis toxin, and indeed no substrates for ADP-ribosylation with ^{3}H-NAD could be detected in polyacrylamide gels of ganglion cell membrane preparations. A large quantity of substrate was seen in brain membranes from the same frogs. Several participants at the symposium expressed the view that the ganglion cells have to have pertussis toxin–sensitive G proteins and that the negative result indicated the presence of a powerful NADase or of an inhibitor of ADP-ribosylation.

include the finding that although a high concentration of GDPβS does blunt the agonist response, it really does not block it as much as in many other systems and it even leads to some irreversible suppression. Also, although a high concentration of GPP(NH)P reduces the reversibility of the agonist response, it is vastly less effective than 1 μM GTPγS in 100 μM GTP. α subunits of several pertussis toxin–insensitive G proteins from mammals have recently been sequenced (Fong et al., 1988; Matsuoka et al., 1988), and it will be interesting to test specific antibodies against these peptides when they become available.

We Could Not Identify I_M in Patches

The next step is to determine how the G protein communicates with M-type K channels. The direct approach would be to record from membrane patches. A simple and important experiment uses on-cell patches. One asks if channels in the patch are modulated when agonist is applied in the bath, suggesting that a diffusible second messenger is involved (e.g., Brum et al., 1984), or if the agonist must be put in the on-cell pipette to be effective, suggesting that whatever carries the signal cannot diffuse far (e.g., Soejima and Noma, 1984). Another basic experiment uses inside-out membrane patches to apply, e.g., kinases, G protein subunits, or other reagents to the cytoplasmic side of the membrane to mimic or block the agonist-induced modulatory signal.

Unfortunately our attempts to record I_M from on-cell or excised patches have been unsuccessful. In on-cell patches we could see a variety of channel openings, but none of them had the appropriate voltage-dependent gating characteristics to be M-type K channels. When large numbers of records from a patch were averaged, we also could not find an averaged component with the gating characteristics of I_M. Several factors might contribute to the difficulty. The channels may be clustered in regions of membranes that are not accessible to probing with patch pipettes in these enzymatically dispersed cells. The channels may have a single-channel current below the noise level of our single-channel records. The process of forming a seal might disrupt M-type K channels or even initiate a local down modulation of I_M. Finally, I_M may be down modulated in all dispersed cells, but some ingredient of our whole-cell recording solution allows I_M to recover in whole-cell experiments.[3]

To estimate the size of the single-channel current, we tried nonstationary, ensemble variance measurements (Sigworth, 1980) using whole-cell recording. Fig. 4 shows the results. The upper trace is the average of 50 whole-cell records. The membrane potential is stepped from −35 to −60 mV and back to induce slow, voltage-dependent closing and reopening of M-type channels. The lower graph shows the point-by-point variance (noisy trace) of the 50 macroscopic records taken pairwise, together with some theoretical expectations. If M-type channels had a single-channel conductance of 3 pS, there would be a kinetic component in the total variance trace with a size and time course that falls somewhere between the two theoretical curves given (see legend). It would be added linearly to the variance contributed by all other channels. Because there does not appear to be an appropriate kinetic component of this size in the data, we conclude that the single-channel conductance is <3 pS. We probably can rule out even a channel of 1.5 pS, which would give a noise transient

---

[3] I_M often increased during the first 2 min after breakthrough into the whole-cell recording configuration.

one-half the size drawn from the model. Using stationary noise analysis during agonist application in other preparations, Marsh and Owen (1989) and Neher et al. (1988) have reported conductances of 1.6 and 3.1 pS for M-type K channels. We could not use stationary noise analysis in our cells because I_M was not the only current modulated by the agonist.

Because of these problems, our electrophysiological work has been confined to whole-cell recording. Despite their small size we do not think that it would be impossible to observe single M-type K channels in normal physiological solutions, but we feel that such a signal might not be robust enough to test second-messenger hypotheses convincingly.

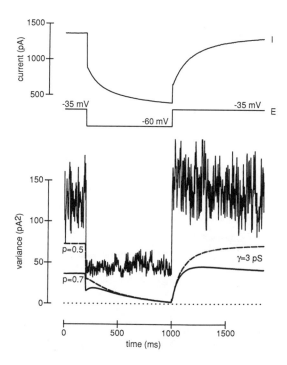

Figure 4. Attempt to resolve noise from gating of M-type K channels. (*Top*) Averaged membrane current during 50 steps from -35 to -60 mV. (*Bottom*) Ensemble variance of 50 current records (noisy trace) and predicted variance for a 3-pS single-channel conductance. The calculation assumes a reversal potential of -100 mV in our standard physiological medium ($[K]_o = 2.5$, $[K]_i = 130$ mM) and gating with the time course of the measured I_M between steady-state open probabilities of 0.5 and 0.1 (*solid line*) or 0.7 and 0.1 (*dashed line*). *Rana catesbeiana*. (Bernheim, L., unpublished).

PLC Is Activated

Do the modulatory agonists activate second-messenger pathways? Muscarinic stimulation has long been known to initiate turnover of membrane phosphoinositides (PI) in mammalian sympathetic ganglia (Hokin, 1966; Bone et al., 1984), and Brown and Adams (1987) have suggested that activation of protein kinase C (PKC) might mediate suppression of I_M in frog ganglia because stimulators of PKC (phorbol esters and diacylglycerols) reduce I_M somewhat. With these clues we decided to look more closely in frog ganglia for activation of PLC and the expected resulting mobilization of intracellular free Ca^{2+} ($[Ca^{2+}]_i$).

PI turnover was evaluated by prelabeling the phospholipids of desheathed sympathetic chains with tritiated inositol and counting the water-soluble inositol compounds released by 1-h incubations in a medium containing modulatory agonists (Pfaffinger et al., 1988). PI turnover was increased 26–81% over background by 0.3 μM SP, 10 μM t-LHRH, or 1 mM muscarine. The stimulation by muscarine was fully blocked by 10

μM atropine. The peak stimulation at individual postganglionic neurons is likely to have been much higher since the experiment used intact sympathetic chains incubated for a long time in potentially desensitizing concentrations of agonist. However, for completeness, the remote possibility that all these agonist effects were on some other cell type in the sympathetic chains should be noted.

Agonist-induced changes of $[Ca^{2+}]_i$ were monitored using the fluorescent Ca^{2+} indicator, fura-2. Single cells were loaded with fura-2 from the recording pipette, and the average $[Ca^{2+}]_i$ was measured under a microscope while the cell was under voltage clamp. Fig. 5 shows that $[Ca^{2+}]_i$ was 30–50 nM in our "resting" conditions and it rose transiently to 65–100 nM during application of t-LHRH. Similar transients were seen with SP and with muscarine (Pfaffinger et al., 1988). In each case the $[Ca^{2+}]_i$ signal returned to baseline within a few minutes even if the agonist remained in the bath. If agonist applications were repeated, the second $[Ca^{2+}]_i$ response was smaller (Fig. 5) and the third was nearly absent. Neither the transience of the original $[Ca^{2+}]_i$ response nor the diminution of subsequent responses were correlated with the desensitization of I_M suppression described earlier. Thus in Fig. 5 the $[Ca^{2+}]_i$ signal is transient but the suppression of I_M response persists until the t-LHRH is removed.

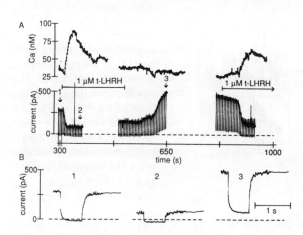

Figure 5. Transient rise of $[Ca^{2+}]_i$ induced by two applications of 1 μM t-LHRH. (*A, top*) Free Ca^{2+} calculated from fura-2 fluorescence signals in a single cell under voltage clamp. (*A, bottom*) Membrane current at -35 mV with test steps to -60 mV in the same cell. Pipette solution included 100 μM fura-2 and 1 mM EGTA with no added Ca^{2+}. (*B*) Selected current records on a faster time scale. (From Pfaffinger et al., 1988.)

We believe that the $[Ca^{2+}]_i$ transient represents mobilization from intracellular stores through the action of inositol 1,4,5-triphosphate (IP_3), but our evidence is incomplete. The lability of the $[Ca^{2+}]_i$ increase is consistent with release from a depletable store. It is not blocked by 5 μM nifedipine, a blocker of L-type Ca channels, but it is occluded when 100 μM IP_3 is in the intracellular pipette solution. Experiments with Ca-free external solution seem ambiguous since removal of external Ca^{2+} not only stops the $[Ca^{2+}]_i$ transient within 1 min but also causes the resting $[Ca^{2+}]_i$ to fall to very low values. We suggest that under these conditions the intracellular stores are emptied rapidly, and that is why there is no agonist-induced transient.

In summary, the three modulatory agonists, muscarine, LHRH, and SP, stimulate PI turnover and evoke a $[Ca^{2+}]_i$ transient that probably comes from intracellular stores (Fig. 2, *right*). In addition, they presumably stimulate PKC through the release of diacylglycerols and possibly release arachidonic acid by the secondary action of diacylglycerol lipase. As any of these might carry a signal required for suppression of I_M, we have tested them all.

Are IP$_3$ or [Ca^{2+}]$_i$ Increases Essential for Suppression of I_M?

IP$_3$

The relevance of IP$_3$ was tested by including 100 μM IP$_3$ in the recording pipette. By itself the added IP$_3$ did not alter I_M, nor did it reduce the ability of agonists to induce suppression of I_M, desensitization, or overrecovery (Pfaffinger et al., 1988). The responses to agonist were also not affected when the pipette contained not IP$_3$ but 290 μM heparin (Bosma, M. M., unpublished observations), which is a competitive IP$_3$ antagonist acting (nonspecifically) at the IP$_3$ receptor with a K_i of 40 nM (Ghosh et al., 1988). These experiments suggest that IP$_3$ plays no essential role as a signal for suppression of I_M.

[Ca^{2+}]$_i$

The relevance of Ca as a signal was tested by artificially raising and lowering [Ca^{2+}]$_i$ (Pfaffinger et al., 1988). Fig. 6 shows that when the pipette contains 5 mM of the powerful Ca^{2+} chelator, BAPTA, the resting [Ca^{2+}]$_i$ falls below 10 nM and that t-LHRH can no longer raise it. Nevertheless, the resting I_M and agonist-induced

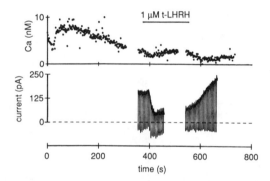

Figure 6. Agonist can suppress I_M even when [Ca^{2+}]$_i$ is strongly buffered. (*Top*) Free Ca^{2+} calculated from fura-2 signals when the pipette solution contains 5 mM BAPTA. (*Bottom*) Membrane current at -35 with steps to -60 mV in the same cell. Zero time is the moment of breakthrough to the whole-cell configuration. (From Pfaffinger et al., 1988.)

suppression and overrecovery remain normal. The responses of I_M were also normal when [Ca^{2+}]$_i$ transients were occluded by IP$_3$ in the pipette. Conversely, the average [Ca^{2+}]$_i$ could be raised above the levels normally achieved with agonist stimulation by repeated depolarizations to membrane potentials that open voltage-gated Ca channels. This also had no effect on the size or responses of I_M. Hence changes of [Ca^{2+}]$_i$ are not the signal for suppression or over recovery of I_M.

Does Activation of PKC Mediate Agonist Actions on I_M?

Brown and Adams (1987) found that the PKC stimulator phorbol dibutyrate (PDBu, 30 μM) partially inhibited I_M, an effect that was not replicated by the "inactive" derivative, 4-α-phorbol. They therefore suggested that activation of PKC may be the normal pathway mediating agonist-induced suppression of I_M. We have confirmed their observations using PDBu, phorbol-12-myristate-13-acetate (PMA), and the synthetic diacylglycerol dioctanoylglycerol (diC$_8$) as stimulators (Pfaffinger et al., 1988; Bosma and Hille, 1989). Fig. 7 A shows that a high concentration of PMA reduces I_M, but not as much as LHRH or SP normally would. The effect does not reverse with washing. Once PMA has been applied, t-LHRH still reversibly suppresses I_M further, and cII-LHRH suppreses I_M and then induces desensitization. The effects of diC_8 were similar to those of PMA except that when diC_8 was washed away, the inhibition of I_M was sometimes partially reversed.

Two of these observations seem to argue against the hypothesis that activation of PKC is the usual signal for LHRH-induced suppression of I_M. First, the reduction of I_M seen with very high concentrations of PKC activators is less than that normally seen with LHRH, and second, while PKC is apparently maximally stimulated, LHRH can still induce further reversible suppression of I_M and desensitization. However, perhaps the pharmacological stimulation of PKC is incomplete because the normal transient increase of $[Ca^{2+}]_i$ is absent. To test this possibility, we raised $[Ca^{2+}]_i$ with a series of depolarizations while PMA was present (Pfaffinger et al., 1988). This did not enhance the action of PMA.

A second effect of PKC activators was evident in our experiments (Leibowitz, M. D., and P. J. Pfaffinger, unpublished observations; Bosma and Hille, 1989). PMA and diC_8 make I_M insensitive to SP. Fig. 7 B shows that SP does nothing to I_M after a treatment with PMA but the action of cII-LHRH is normal, as in Fig. 7 A. The same effect is produced by diC_8 if SP is applied before the diC_8 is washed off. In our

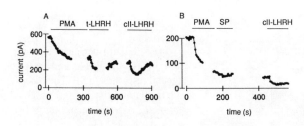

Figure 7. Stimulation of PKC by phorbol ester depresses I_M and blocks the response to SP but not to LHRH. Membrane current at -35 mV is plotted as circles as 1 μM PMA, SP, t-LHRH, and cII-LHRH are applied. (From Bosma and Hille, 1989.)

experiments using I_M as an indicator, this effect of PKC activators resembles the desensitization induced by a high concentration of SP.

Having two effects of PKC activators gave us a handle on testing whether PKC normally carries the signal. The strategy was to apply inhibitors of PKC, to prove that they actually are inhibiting PKC by showing that they block the two effects of PKC activators, and then to see if the actions of agonists on I_M are prevented by the inhibitors. We used three inhibitors. H-7, a nonspecific kinase inhibitor, was applied in the bath at 300 μM. Staurosporine, a potent inhibitor of PKC with an IC_{50} of 3 nM, and of other kinases at slightly higher concentrations (Tamaoki et al., 1986), was used in the pipette at 100–200 nM. PKC(19-36), a specific pseudosubstrate peptide with an IC_{50} of 0.2 μM for block of PKC in a cell-free system (House and Kemp, 1987), was applied in the pipette at 1.5–3 μM. Results with staurosporine are shown in Fig. 8. The tests were begun 20 min after breakthrough to allow staurosporine time to enter the cell. In this experiment the two effects of PMA have been blocked by staurosporine since PMA neither reduces I_M nor disturbs the normal sequence of suppression, desensitization, and overrecovery with SP. We conclude that PKC has been strongly inhibited and that even with PKC inhibited, SP is fully active. Staurosporine also blocks the actions of diC_8. The results with the other two inhibitors, nonspecific H-7 and highly specific PKC(19-36), are similar except that the inhibition seems not to be as strong since PMA still decreases I_M a little. This experiment argues that activation of PKC is not needed for normal agonist-induced suppression of I_M or for desensitization or overrecovery.

The lack of specificity of H-7 and staurosporine is no disadvantage when the outcome is negative, as here. Indeed it could be exploited to rule out the importance of any other H-7 and staurosporine-sensitive kinases in a single test.

Hydrolyzable ATP Is Required to Maintain I_M

When we first started whole-cell clamping of sympathetic cells, we had to determine what pipette (intracellular) solutions gave the most robust I_M. A striking result was the need for intracellular ATP. If a pipette solution without ATP is used, I_M is maintained for <2 min and then falls to nearly zero within a few more minutes (Pfaffinger, 1988). The nonhydrolyzable ATP analogue, 5'-adenylylimidodiphosphate [APP(NH)P], would not substitute for intracellular ATP. These observations suggest that phosphorylation of some intracellular component must be continuously maintained for I_M to survive. A simple hypothesis that we have not tested would be that the reversible suppression of I_M by agonists involves a cycle of dephosphorylation and of rephosphorylation of this component.

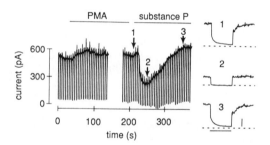

Figure 8. Staurosporine blocks depression of I_M by PMA and preserves normal responses to SP. The recording starts 20 min after breakthrough with a pipette solution including 100 nM staurosporine. The bath is first switched to one containing 1 µM PMA and subsequently, to 1 µM SP. Scale bars for insets at right: 200 pA, 600 ms. (From Bosma and Hille, 1989.)

Do Arachidonic Acid Metabolites Carry the Message?

Arachidonic acid (AA) can be released in at least two ways, directly by the receptor-mediated activation of PLA_2 and secondarily by the action of diacylglycerol lipase on diacylglycerols. As modulatory agonists activate PLC and probably release diacylglycerols, they may also release arachidonic acid. Biochemical assays for agonist-stimulated release of AA and its metabolites have not been done in frog sympathetic ganglia.

We tested the effects of adding 100 µM AA to the bath. It reversibly increased I_M in seven out of seven cells (Fig. 9), which suggested the hypothesis that AA metabolites might contribute to the maintenance, recovery, or overrecovery of I_M rather than to its suppression. Consistent with this suggestion, two nonspecific blockers of AA production, bromphenacyl bromide (BPAB, 200 µM) and quinacrine (25 or 50 µM), reversibly reduced I_M, and a lower concentration of BPAB (20 µM), made the actions of t-LHRH (but not SP) irreversible. AA did not affect the ability of SP or LHRH to suppress I_M or to give desensitization and overrecovery. These observations are intriguing, but more specific methods are required to test the hypothesis.

What Is Desensitization?

When high concentrations of muscarine, LHRH, or SP are applied to the bath, I_M is first suppressed and then begins to recover spontaneously (Jones, 1985; Pfaffinger, 1988; Bosma and Hille, 1989). The rate of this desensitization can be measured as the percentage of the suppressed I_M returning per minute. Desensitization is not obvious unless one applies agonist concentrations well above those needed to suppress I_M. For example, 10–30 nM SP or cII-LHRH suppress I_M well with little desensitization, and 1 μM desensitizes at ~50%/min; on the other hand, 300 nM t-LHRH is needed to suppress I_M well, and 1 μM desensitizes at only 7%/min; similarly >1 μM muscarine is needed to suppress I_M well in our dialyzed cells and 20 μM desensitizes at only 5–10%/min.

This desensitization is homologous and long lasting. cII-LHRH desensitizes responses to cII- and t-LHRH but not to SP or muscarine, and SP desensitizes

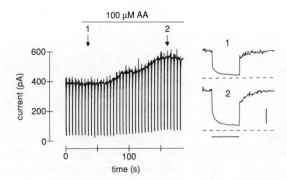

Figure 9. AA enhances I_M. (*Top*) continuous record of membrane current before and during addition of 100 μM AA to the bath. (*Bottom*) Current responses during steps from −35 to −60 mV. Bars = 200 pA and 600 ms. *Rana catesbeiana*. (Bosma, M. M., unpublished.)

responses to SP and not to LHRH. There is no recovery from desensitization in the 30–40 min that we can continue to record from the same cell.

It would be interesting to know the mechanism of homologous desensitization since, at least in whole-cell recording, it is a form of long-lasting and specific synaptic modulation. Desensitization is not prevented when cells are first incubated with 8.5 μM concanavalin A (unpublished observations), a lectin that in submicromolar concentrations prevents the rapid desensitization of non–N-methyl-D-aspartate (NMDA) glutamate receptors (Mathers and Usherwood, 1978; Mayer and Vyklicky, 1989). Our experiments rule out the involvement of Ca^{2+}, IP_3, and AA, as these second messengers do not mimic or prevent desensitization. Activation of PKC does mimic desensitization to a degree but differs from the full agonist-induced effect in at least two respects. First, it is confined to SP receptors, and strong PKC activation by phorbol esters does not accelerate the slow desensitization seen with t-LHRH or impede the fast desensitization seen with cII-LHRH. Second, when induced by reversible stimulators of PKC (e.g., diC_8) the loss of response to SP is itself readily reversed by washing. Thus the PKC-induced "desensitization" of SP receptors is not long lasting. Moreover, as each of the agonists seem to activate the PLC messenger system, it would be hard to explain the homologous nature of desensitization by invoking components of that system.

A highly plausible mechanism for homologous desensitization is that proposed originally for the β-adrenergic receptor (Benovic et al., 1986; also see chapter 7 by Benovic et al. in this volume). A protein kinase, β-adrenergic receptor kinase (βARK), has been identified that phosphorylates several kinds of receptors (β, α, muscarinic, rhodopsin) but only when they are occupied by agonists. Such substrate specificity guarantees that only homologous modification occurs even though the enzyme is not specific for one receptor type. Lohse et al. (1989) have shown that inhibition of this enzyme by heparin ($K_i = 6$–20 nM) or by a peptide containing amino acids 57–70 of the β_2-adrenergic receptor, will block agonist-induced phosphorylation and homologous desensitization of β-adrenergic receptors. We have found that desensitization of SP and LHRH responses is not prevented by including 290 μM heparin or 600 μM of the peptide used by Lohse et al. in the recording pipette solution (unpublished observations). Therefore, homologous desensitization with LHRH and SP is not mediated by the classical βARK.

Overview

Attempts to understand the signaling system(s) underlying control of I_M have been difficult. Our results in frog sympathetic ganglia are summarized in Fig. 2, *right*. A pertussis toxin–insensitive G protein couples the muscarinic and LHRH receptors to suppression of I_M. The three receptors (muscarinic, LHRH, and SP) can activate PLC as evidenced by PI turnover and transients of $[Ca^{2+}]_i$. Our experiments did not directly ask if PLC was being activated by the pertussis toxin–insensitive G protein, but in other systems this is usually true, see chapter 5 by Morris et al. (Harden) in this volume. Nevertheless, we could show that increases of IP_3 or $[Ca^{2+}]_i$ or activation of PKC are not the signals that normally suppress I_M or induce homologous desensitization of muscarinic, LHRH, or SP receptors. Two inhibitors of βARK, a kinase that is thought to underlie homologous desensitization of other G protein–coupled receptors, failed to block homologous desensitization with SP or LHRH.

We have yet to test directly whether muscarine, LHRH, and SP activate PKC. Pharmacological activation of PKC can block the actions of SP, but the effects does not fully mimic agonist-induced desensitization of that pathway. PKC can also reduce M currents, but again the mimicry of agonist-induced suppression is incomplete and inhibition of PKC does not interfere with the actions of LHRH or SP. Judging from the known properties of other G protein-coupled receptors and voltage-gated ion channels, both the SP receptor and M-type K channels may have cytoplasmic domains with serine and threonine residues capable of phosphorylation by PKC, which, however, would not be the principle site of action of the agonist actions we are studying.

The mechanism of regulation of I_M has been studied in several other laboratories. The groups of Adams and Brown report that I_M is reduced by pharmacological stimulators of PKC but not by injected IP_3, Ca^{2+}, or cyclic nucleotides in frog and rat sympathetic ganglia and in the neuroblastoma x glioma cell line, NG108-15 (Adams et al., 1982; Higashida and Brown, 1986; Brown and Adams, 1987; Brown and Higashida, 1988; Brown et al., 1989). Muscarinic suppression of I_M is inhibited by GDPβS, made irreversible by GTPγS, and is not pertussis toxin sensitive in rat superior cervical ganglion cells (Brown et al., 1989). In hippocampal pyramidal cells the coupling of I_M

to muscarinic receptors is also not pertussis toxin sensitive (Dutar and Nicoll, 1988). Our observations agree with all of these. On the other hand, phorbol esters do not decrease I_M in hippocampal pyramidal cells and intracellular IP_3 does (Malenka et al., 1986; Dutar and Nicoll, 1988). In addition, β-adrenergic agonists, forskolin, and cAMP analogues augment I_M in smooth muscle and somatostatin augments I_M in neurons of the solitary tract (Jacquin et al., 1988; Sims et al., 1988). These observations differ from those in sympathetic neurons.

After many negative results and apparent differences among pathways, what useful picture remains? The finding that three agonists activate PLC will ultimately help us to understand some responses of sympathetic neurons. The finding that nevertheless neither Ca^{2+} nor IP_3 mediates the agonist-induced actions on I_M and that PKC is at least not essential reminds us that G protein–coupled pathways branch. There is a richness of responses to one agonist, so finding stimulation of one pathway does not prove its involvement in a particular effect. Similarly, the finding that PKC stimulators reduce I_M reminds us that mimicry alone does not prove a pathway and suggests that there exist other pathways that may regulate I_M via PKC. The hypothesis that I_M is suppressed by the direct actions of an activated, pertussis toxin–insensitive G protein (without other second messengers) remains a viable hypothesis in all cells where I_M has been found. As Yatani and Brown (1989) have suggested for β-adrenergic modulation of Ca channels in heart, there could be a fast pathway for modulating I_M via a membrane-delimited, direct G protein action and slower parallel pathways involving cytoplasmic second messengers. Alternatively, the requirement for hydrolyzable ATP would be consistent with a mechanism based on a novel G protein–coupled phosphatase. Finally, the hypothesis that suppression involves a hitherto undescribed diffusible messenger is also viable. The greatest progress will be made if a way to record I_M in excised patches can be developed.

Acknowledgments

We are deeply grateful to Neil Nathanson, who introduced us to G protein–coupled receptors and who has collaborated on several of these projects and offered good advice throughout. We thank Wolfhard Almers for collaboration on the $[Ca^{2+}]_i$ measurements and Emily Subers for collaboration in the measurement of PI turnover. Don Anderson built apparatus and Lea Miller typed the manuscript.

This work was supported by a National Institutes of Health research grant NS-08174, a Research Award from the McKnight Endowment Fund for Neuroscience, and stipend support from the National Institutes of Health training grants GM-07270 and NS-07097, the Muscular Dystrophy Association, and the Fonds National Suisse de la Recherche Scientifique. M. D. Leibowitz was a Fondation pour l' Etude du Système Nerveux Central et Périphérique (FESN) scholar in "Transduction of Neuronal Signals."

References

Adams, P. R., and D. A. Brown, 1980. Luteinizing hormone-releasing factor and muscarinic agonists act on the same voltage-sensitive K^+-current in bullfrog sympathetic neurons. *British Journal of Pharmacology*. 68:353–355.

Adams, P. R., D. A. Brown, and A. Constanti. 1982. Pharmacological inhibition of the M-current. *Journal of Physiology*. 332:223–262.

Adams, P. R., D. A. Brown, and S. W. Jones. 1983. Substance P inhibits the M-current in bullfrog sympathetic neurones. *British Journal of Pharmacology.* 79:330–333.

Benovic, J. L., F. Mayor, Jr., R. L. Somers, M. G. Caron, and R. J. Lefkowitz. 1986. Light-dependent phosphorylation of rhodopsin by β-adrenergic receptor kinase. *Nature.* 321:869–872.

Bley, K. R., and R. W. Tsien. 1988. LHRH and substance P inhibit N- and L-type calcium channels in frog sympathetic neurons. *Biophysical Journal.* 53:253a. (Abstr.)

Bone, E. A., P. Fretten, S. Palmer, C. J. Kirk, and R. H. Michell. 1984. Rapid accumulation of inositol phosphates in isolated rat superior cervical sympathetic ganglia exposed to V_1 vasopressin and muscarinic cholinergic stimuli. *Biochemical Journal.* 221:803–811.

Bosma, M. M., and B. Hille. 1989. Protein kinase C is not necessary for peptide-induced suppression of M-current or for desensitization of the peptide receptors. *Proceedings of the National Academy of Sciences.* 86:2943–2947.

Breitwieser, G. E., and G. Szabo. 1988. Mechanism of muscarinic receptor-induced K^+ channel activation as revealed by hydrolysis-resistant GTP analogues. *Journal of General Physiology.* 91:469–493.

Brown, D. A. 1988. M currents. *In* Ion Channels. T. Narahashi, editor. Plenum Publishing Corp., New York. 1:55–94.

Brown, D. A., and P. R. Adams. 1980. Muscarinic suppression of a novel voltage-sensitive K^+-current in a vertebrate neurone. *Nature.* 283:673–676.

Brown, D. A., and P. R. Adams. 1987. Effects of phorbol dibutyrate on M currents and M current inhibition in bullfrog sympathetic neurons. *Cellular and Molecular Neurobiology.* 7:255–269.

Brown, D. A., and H. Higashida. 1988. Membrane current responses of NG108-15 mouse neuroblastoma x rat glioma hybrid cells to bradykinin. *Journal of Physiology.* 397:167–184.

Brown, D. A., N. V. Marrion, and T. G. Smart. 1989. On the transduction mechanism for muscarine-induced inhibition of M-current in cultured rat sympathetic neurones. *Journal of Physiology.* 413:469–488.

Brown, D. A., and A. A. Selyanko. 1985. Two components of muscarine-sensitive membrane current in rat sympathetic neurones. *Journal of Physiology.* 358:335–363.

Brum, G., W. Osterrieder, and W. Trautwein. 1984. β-Adrenergic increase in the calcium conductance of cardiac myocytes studied with the patch clamp. *Pflügers Archiv.* 401:111–118.

Dutar, P., and R. A. Nicoll. 1988. Classification of muscarinic responses in hippocampus in terms of receptor subtypes and second-messenger systems: electrophysiological studies in vitro. *Journal of Neuroscience.* 8:4214–4224.

Evans, M. G., and A. Marty. 1986. Calcium dependent chloride currents in isolated cells from rat lacrimal glands. *Journal of Physiology.* 378:437–460.

Fong, H. K., K. K. Yoshimoto, P. Eversole-Cire, and M. I. Simon. 1988. Identification of a GTP-binding protein α subunit that lacks an apparent ADP-ribosylation site for pertussis toxin. *Proceedings of the National Academy of Sciences.* 85:3066–3070.

Ghosh, T. K., P. S. Eis, J. M. Mullaney, C. L. Ebert, and D. L. Gill. 1988. Competitive, reversible, and potent antagonism of inositol 1,4,5-trisphosphate-activated calcium release by heparin. *Journal of Biological Chemistry.* 263:11075–11079.

Higashida, H., and D. A. Brown. 1986. Two polyphosphatidylinositide metabolites control two K^+ currents in a neuronal cell. *Nature*. 323:333–335.

Hokin, L. E. 1966. Effects of acetylcholine on the incorporation of ^{32}P into various phospholipids in slices of normal and denervated superior cervical ganglia of the cat. *Journal of Neurochemistry*. 13:179–184.

House, C., and B. E. Kemp. 1987. Protein kinase C contains a pseudosubstrate prototope in its regulatory domain. *Science*. 238:1726–1728.

Jacquin, T., J. Champagnat, S. Madamba, M. Denavit-Saubié, and G. R. Siggins. 1988. Somatostatin depresses excitability in neurons of the solitary tract complex through hyperpolarization and augmentation of I_M, a non-inactivating voltage-dependent outward current blocked by muscarinic agonists. *Proceedings of the National Academy of Sciences*. 85:948–952.

Jan, L. Y., and Y. N. Jan. 1982. Peptidergic transmission in sympathetic ganglia of the frog. *Journal of Physiology*. 327:219–246.

Jones, S. W. 1985. Muscarinic and peptidergic excitation of bull-frog sympathetic neurones. *Journal of Physiology*. 366:63–87.

Jones, S. W., and T. N. Marks. 1989. Calcium currents in bullfrog sympathetic neurons. I. Activation kinetics and pharmacology. *Journal of General Physiology*. 94:151–167.

Kuba, K., and K. Koketsu. 1976. Analysis of the slow excitatory postsynaptic potential in bullfrog sympathetic ganglio cells. *Japanese Journal of Physiology*. 26:651–669.

Kuffler, S. W., and T. J. Sejnowski. 1983. Peptidergic and muscarinic excitation at amphibian synapses. *Journal of Physiology*. 341:257–278.

Kurachi, Y., T. Nakajima, and T. Sugimoto. 1986. On the mechanism of activation of muscarinic K^+ channels by adenosine in isolated atrial cells: involvement of GTP-binding proteins. *Pflügers Archiv*. 407:264–274.

Lohse, M. J., R. J. Lefkowitz, M. G. Caron, and J. L. Benovic. 1989. Inhibition of β-adrenergic receptor kinase prevents rapid homologous desensitization of $β_2$-adrenergic receptors. *Proceedings of the National Academy of Sciences*. 86:3011–3015.

Malenka, R. C., D. V. Madison, R. Andrade, and R. A. Nicoll. 1986. Phorbol esters mimic some cholinergic actions in hippocampal pyramidal neurons. *Journal of Neuroscience*. 6:475–480.

Marsh, S. J., and D. G. Owen. 1989. Fluctuation analysis of M-current in cultured rat sympathetic neurones. *Journal of Physiology*. 410:31P.

Mathers, D. A., and P. N. R. Usherwood. 1978. Effects of concanavalin A on junctional and extrajunctional L-glutamate receptors on locust skeletal muscle fibres. *Comparative Biochemistry and Physiology*. 59C:151–155.

Matsuoka, M., H. Itoh, T. Kozasa, and Y. Kaziro. 1988. Sequence analysis of cDNA and genomic DNA for a putative pertussis toxin-insensitive guanine nucleotide-binding regulatory protein α subunit. *Proceedings of the National Academy of Sciences*. 815:5384–5388.

Mayer, M. L., and L. Vyklicky, Jr. 1989. Concanavalin A selectively reduces desensitization of mammalian neuronal quisqualate receptors. *Proceedings of the National Academy of Sciences*. 86:1411–1415.

Nathanson, N. M. 1987. Molecular properties of the muscarinic acetylcholine receptor. *Annual Review of Neuroscience*. 10:195–236.

Neher, E., A. Marty, K. Fukuda, T. Kubo, and S. Numa. 1988. Intracellular calcium release mediated by two muscarinic receptor subtypes. *FEBS Letters*. 240:88–94.

Otera, A. S., G. F. Breitwieser, G. Szabo. 1988. Activation of muscarinic potassium currents by ATPγS in atrial cells. *Science.* 242:443–445.

Pfaffinger, P. J. 1988. Muscarine and t-LHRH suppress M-current by activating an IAP-insensitive G-protein. *Journal of Neuroscience.* 8:3343–3353.

Pfaffinger, P. J., M. D. Leibowitz, E. M. Subers, N. M. Nathanson, W. Almers, and B. Hille. 1988. Agonists that suppress M-current elicit phosphoinositide turnover and Ca^{2+} transients, but these events do not explain M-current suppression. *Neuron.* 1:477–484.

Sigworth, F. J. 1980. The variance of sodium current fluctuations at the node of Ranvier. *Journal of Physiology.* 307:97–129.

Sims, S. M., J. J. Singer, and J. V. Walsh, Jr. 1988. Antagonistic adrenergic-muscarinic regulation of M current in smooth muscle cells. *Science.* 239:190–193.

Soejima, M., and A. Noma. 1984. Mode of regulation of the ACh-sensitive K-channel by the muscarinic receptor in rabbit atrial cells. *Pflügers Archiv.* 400:424–431.

Sternweis, P. C., and A. G. Gilman. 1982. Aluminum: a requirement for activation of the regulatory component of adenylate cyclase by fluoride. *Proceedings of the National Academy of Sciences.* 79:4888–4891.

Tamaoki, T., H. Nomoto, I. Takahashi, Y. Kato, M. Morimoto, and F. Tomita. 1986. Staurosporine, a potent inhibitor of phospholipid/Ca^{2+} dependent protein kinase. *Biochemical and Biophysical Research Communications.* 135:397–402.

Yatani, A., and A. M. Brown. 1989. Rapid β-adrenergic modulation of cardiac calcium channel currents by a fast G protein pathway. *Science.* 245:71–74.

Chapter 5

Regulation of Phospholipase C

Andrew J. Morris, Gary L. Waldo, Jose L. Boyer,
John R. Hepler, C. Peter Downes, and T. Kendall Harden

*Department of Pharmacology, University of North Carolina School
of Medicine, Chapel Hill, North Carolina 27599*

Introduction

Phospholipase C (PLC) catalyzes the hydrolysis of phosphatidylinositol(4,5) bisphosphate [PtdIns(4,5)P_2] to form the second messengers inositol (1,4,5)P_3 and diacylglycerol in response to multifarious hormones, neurotransmitters, growth factors, antigens and chemoattractants (Downes and Michell, 1985; Berridge, 1987). Whereas considerable and specific knowledge of the production, action, and metabolism of the second messengers of this major signaling system has been accrued in recent years, unambiguous identity of the proteins involved in PLC–associated transmembrane signaling and a description of their mechanisms of purposeful interaction remain to be accomplished. The goal of this review is to assimilate some of the salient information concerning these signaling proteins, and to summarize the value of the turkey erythrocyte as a model system to study a hormone- and putative guanine nucleotide regulatory protein-activated PLC.

PLC

PLC Isozymes

Phosphoinositide-specific PLC has been purified to homogeneity by a number of laboratories from a variety of tissues (for reviews see Boyer et al., 1989b; Crooke and Bennett, 1989; Rhee et al., 1989). Purification strategies have most often used a cytosolic fraction of enzyme as starting material. The enzymes have been assayed during purification using either phosphatidylinositol or PtdIns(4,5)P_2 as substrate presented as components of either small unilamellar phospholipid vesicles or of mixed phospholipid and detergent micelles. Under these assay conditions, PLC activity is generally maximal at supraphysiological concentrations of Ca^{++}. Molecular masses ranging from 56 to 154 kD have been reported for these proteins (see Rhee et al., 1989 for review). The work of Rhee and coworkers with monoclonal antibodies against enzymes purified from bovine brain has greatly facilitated the delineation of what clearly is a multiplicity of PLC isozymes (Suh et al., 1988a; Rhee et al., 1989). Their systematic consideration of the immunological properties of the enzymes purified to date has led to the recent proposal of a nomenclature for the PLC isozymes that is not based on the unpredictabilities of assignment of protein differences based on estimation of apparent molecular mass (Rhee et al., 1989). At a minimum, they propose that at least five (α [56 kD], β [154 kD], γ [148 kD], δ [85 kD], and ϵ [85 kD] isozymes exist. Molecular cloning of cDNAs encoding PLC-α (Bennett et al., 1988), -β (Katan et al., 1988; Suh et al., 1988b), -γ (Stahl et al., 1988; Suh et al., 1988c), and -δ (Suh et al., 1988b) has been accomplished. The predicted amino acid sequence of these isozymes encodes proteins with M_r values of 56, 138–139, 148, and 85 kD, respectively.

Overall amino acid homology of the proteins is low. However, PLC-β, -γ, and -δ contain two highly conserved sequences of ~150 and 120 amino acids. These domains, designated X and Y, have no established function, although it can be speculated that they may be involved in catalysis. The remaining sequence of these proteins is quite diverse, perhaps hinting at differential modes of regulation. A surprise emanating from the initial cloning of the PLC-γ isozyme is that it shares three regions (designated A, B, and C) of sequence homology with noncatalytic domains of both the src family of proteins and the newly discovered viral crk oncogene (Stahl et al., 1988; Suh et al., 1988c; Rhee et al., 1989). Again, function is not established, but purpose in interaction with regulatory proteins is an obvious possibility.

Recent data suggest that these identified isozymes of PLC may form the prototype members of several related subgroups of PLC isozymes. This evidence has come both from attempts to purify new PLC isozymes and from the use of molecular genetic approaches to search for DNA sequences encoding PLC isozymes. Three isozymes of PLC of 61, 63, and 69 kD have been purified from human platelet membranes (Banno et al., 1988; Baldassare et al., 1989). In addition, an isozyme of PLC of 57 kD was purified from platelet cytosol, and antibodies raised against this protein did not react with the 63- and 69-kD membrane-derived PLCs (Baldassare et al., 1989). Although the apparent size of these proteins are similar to that of PLC-α their relationship to this isozyme of PLC has not yet been established. An isozyme of PLC with a molecular weight of 85 kD has been purified from bovine brain and a partial amino acid sequence derived from this protein differs from that of PLC-δ (Meldrum et al., 1989). Perhaps this isozyme of PLC represents the bovine counterpart of PLC-ϵ originally purified from rat brain (Homma et al., 1988). Another protein with potential yet unexplored similarity to PLC-δ and -ϵ with a molecular weight of 87 kD has been purified from rat liver (Fukui et al., 1988). Two isozymes of PLC with a possible relationship to PLC-β and -γ have been purified from bovine platelets (Hakata et al., 1982) and turkey erythrocytes (Morris et al., 1990a). The bovine platelet PLC was reported to have a molecular weight of 143 kD. The turkey erythrocyte protein has an apparent size of 150 kD and does not react with mixtures of monoclonal antibodies raised against bovine brain PLC-β and -γ (see page 70). The molecular genetic approach has revealed two further PLC isomzymes. A sequence has been identified in a human lymphocyte cDNA library potentially encoding a PLC with the X and Y and A, B, and C regions of PLC-γ, but with little sequence homology elsewhere (Ohta et al., 1989). The genetic lesion of the *Drosophila* visual mutant norpA involves mutation of a gene encoding a protein with sequence similarity to that of PLC-β (Bloomquist et al., 1988). Although several of the identified PLC isozymes were initially purified from tissue sources composed of heterogeneous cell types it is now clear that multiple isozymes of PLC can coexist in a single cell type (Meisenhelder et al., 1989; Wahl et al., 1989), providing further evidence that they may serve different cellular functions. Table I summarizes current knowledge of the interrelationships between the identified isozymes of PLC and tentatively attempts to incorporate the as yet unclassified PLC isozymes described above into this scheme.

Regulation of PLC

Little unambiguous information on the regulation of PLC is available. In the case of hormonal regulation of phosphoinositide signaling, the subcellular localization of the enzyme must be reconciled with the realities of an initial step involving activation of a cell surface receptor and with consideration of a terminal step that involves action on substrates, which are themselves constituents of cellular membrane structures (predominantly the plasma membrane). Most of the PLC isozymes purified to date have come from soluble tissue extracts. These isozymes can be extracted from cells and purified as stable proteins without exposure to detergents. Moreover, the deduced primary sequences of the four cloned PLC isozymes do not contain regions of sufficient hydrophobicity to serve according to generally accepted criteria as membrane anchors. Attempts to specifically purify a membrane-associated isozyme of PLC from bovine brain resulted in the identification of a protein (PLC-β) previously purified from a soluble extract of the same tissue (Lee et al., 1987). Significant extraction of PLC-β from bovine brain membranes could be effected by washing them with high concentrations of KCl, supporting the idea that membrane association of PLC does not require membrane insertion. There is also immunological evidence that PLC-α and -γ can exist as membrane-associated forms (Lee et al., 1987; Bennett and Crooke, 1987). An

TABLE I
Summary of Properties of PLC Isozymes

Nomenclature of Rhee et al., 1989	Source	Molecular weight		X-Y region sequence homology	Sequence homology to src family nonreceptor tyrosine kinases	Related PLC isozymes (purified proteins or putative PLCs identified by cDNA cloning)
		SDS-PAGE	cDNA			
		kD	kD			
PLC-α	Rat liver	68[1]				
	Sheep seminal vesicles	65[2]		Poor	No	69, 63[13], and 64[14] kD human platelet membrane PLCs
	Guinea pig uterus	62[3]	56.6[9]			57-kD human platelet cytosolic PLC[13]
PLC-β	Bovine brain membranes	154[4]	138.2[10]	Good	No	*Drosophila* norpA gene product[15]
						143-kD bovine platelet PLC[16]
	Bovine brain	150 (140,100)[5]	138.6[11]			150-kD turkey erythrocyte PLC[17]
PLC-γ	Bovine brain	145[5]	148.4[11,12]	Good	Yes	146.1-kD PLC encoded by human lymphocyte–derived cDNA[18]
						143-kD bovine platelet PLC[16]
PLC-δ	Bovine brain	85,[6] 88[7]	85.8[11]	Good	No	87-kD rat liver PLC[19]
	Rat brain	85[8]				
PLC-ξ	Rat brain	85[8]	Unknown	Unknown	Unknown	87-kD rat liver PLC[19]
						85-kD bovine brain PLC[20]

1. Takenawa, T., and Y. Nagai. 1981. *J. Biol. Chem.* 256:6769–6775.
2. Hoffmann, S. L., and P. W. Majerus. 1984. *J. Biol. Chem.* 257:6461–6469.
3. Bennett, C. F., and S. T. Crooke. 1987. *J. Biol. Chem.* 262:13789–13797.
4. Katan, M., and P. J. Parker. 1987. *Eur. J. Biochem.* 168:413–418.
5. Ryu, S. H., K. S. Cho, K. Y. Lee, P. G. Suh, and S. G. Rhee. 1987. *J. Biol. Chem.* 262:12511–12518.
6. Ryu, S. H., P. G. Suh, K. S. Cho, K. Y. Le, and S. G. Rhee. 1987. *Proc. Natl. Acad. Sci. (USA).* 84:6649–6653.
7. Rebecchi, M. J., and O. M. Rosen. 1987. *J. Biol. Chem.* 262:12526–12532.
8. Homma, Y., J. Imaki, O. Nakonishi, and T. Takenawa. 1988. *J. Biol. Chem.* 263:6592–6598.
9. Bennett, C. F., J. M. Balcarek, A. Varrichio, and S. T. Crooke. 1988. *Nature.* 334:268–271.
10. Katan, M., R. W. Kriz, N. Totty, R. Philip, E. Meldrum, R. A. Aldape, J. L. Knopf, and P. J. Parker. 1988. *Cell.* 54:171–177.
11. Suh, P. G., S. H. Ryu, K. H. Moon, M. W. Suh, and S. G. Rhee. 1988. *Cell.* 54:161–169.
12. Stahl, M. L., C. R. Ferenz, K. L. Kelleher, R. W. Kriz, and J. L. Knopf. 1988. *Nature.* 332:269–272.
13. Baldassare, J. J., P. A. Henderson, and G. J. Fisher. 1989. *Biochemistry.* 28:6010–6016.
14. Banno, Y., Y. Yada, and Y. Nozawa. 1988. *J. Biol. Chem.* 263:11459–11464.
15. Bloomquist, B. T., R. D. Shortridge, S. Schnewly, M. Perdew, C. Montell, H. Steller, G. Rubin, and W. L. Pak. 1988. *Cell.* 54:723–733.
16. Hakata, H., J.-L. Kambayashi, and G. Kosaki. 1982. *J. Biochem.* 92:929–935.
17. Morris, A. J., G. L. Waldo, C. P. Downes, and T. K. Harden. 1990. *J. Biol. Chem.* In press.
18. Ohta, S., A. Matsui, Y. Nazawa, and Y. Kagawa. 1988. *FEBS Lett.* 242:31–35.
19. Fukui, T., R. J. Lutz, and J. M. Lowenstein. 1988. *J. Biol. Chem.* 263:17730–17737.
20. Meldrum, E., M. Katan, and P. J. Parker. 1989. *Eur. J. Biochem.* 182:673–677.

isozyme of PLC ($M_r = 62$ kD) has been purified from deoxycholate extracts of human platelet membranes (Baldassare et al., 1989). This protein appears to be distinct from a purified cytosolic form of PLC derived from the same tissue. However, it is not clear if this PLC is exclusively membrane associated in platelets, and its relationship to the other identified isozymes of PLC remains to be established.

The mechanism(s) by which PLC associates with cellular membranes remains speculative. Specific association with protein and/or lipid components of the membrane is an obvious possibility, or a posttranslational modification such as fatty acylation might facilitate membrane interaction. It should be noted that interaction of purified PLC with phospholipid vesicles has been found to be substrate dependent (Hofmann and Majerus, 1982; Herrero et al., 1988). As is discussed below (page 66), a potential capacity of PLC, i.e., PLC-γ, to be variably associated among subcellular localizations in response to growth factors, could form an important component of regulation of tyrosine kinase receptor signaling systems.

Various comparisons have been made among the purified enzymes regarding their absolute catalytic activities, their dependence on Ca^{++} for catalysis, and their relative selectivites for the three potential phosphoinositide substrates. Although these comparisons may be instructive in the classification of PLC isozymes, it is important to stress that the catalytic activity of phospholipases (including PLC) is known to be affected profoundly by the phospholipid composition and physicochemical state of the substrate preparation (Irvine et al., 1979a, b; Hoffmann and Majerus, 1982; Herrero et al., 1988). Furthermore, although the mechanism is not known, PLC activity directed against vesicular or micellar substrates greatly exceeds that against substrates presented in an artificial phospholipid monolayer (Irvine et al., 1979a, b; Hirasawa et al., 1981). Little work has been reported concerning the activity of purified PLC isozymes with physiologically relevant substrate preparations in which the substrates are components of a protein-containing phospholipid bilayer. We have noted (page 70) that a PLC purified from turkey erythrocytes shows substantial activity against mixed micellar substrate preparations, but is inactive when recombined in the absence of hormonal activators with turkey erythrocyte ghosts containing endogenous radiolabeled substrates.

On the basis of the high specific activity of the purified brain enzymes relative to the amount of PtdIns(4,5)P_2 in brain membranes, Rhee and co-workers (1989) have proposed that a negative modulator of PLC is important in its physiological regulation, and that PLC could be activated in response to a neurotransmitter as a consequence of the agonist-occupied receptor in some way relieving an inhibitory constraint. Perhaps such a mechanism could explain the existence of sequence similarity of PLC-γ to regulatory regions of nonreceptor tyrosine kinases (Stahl et al., 1988; Suh et al., 1988b). There are no reports that directly implicate such a mechanism for activation of PLC, although several studies suggest that hormone receptors may inhibit PLC (see Linden and Delahunty, 1989), possibly through the action of a G protein, e.g., a protein analogous to Gi in the adenylate cyclase regulatory cycle.

G Proteins and PLC

Stimulated initially from the ideas outlined by Michell (1975), it became clear that members of many receptor classes shared in common a capacity to mobilize Ca^{++} and to stimulate phosphoinositide hydrolysis. Likewise, agonist binding to a number of these PLC-activating/Ca^{++}-mobilizing receptors subsequently was shown to be regulated by guanine nucleotides (see Harden, 1989). The potential role of G protein(s) in phosphoinositide/Ca^{++} signaling came clearly into focus as a result of reports in 1985 by Litosch et al. (1985) and Cockcroft and Gomperts (1985). These and other workers (see Harden, 1989; Martin, 1989) demonstrated that stable analogues of GTP activate PLC in membranes prepared from a variety of tissues, and that in many cases guanine nucleotide-dependent activation of the enzyme by hormones, neurotransmitters, or chemoattractants could be observed. Progress past observation of the general phenomenology of guanine nucleotide dependence for activation of PLC by hormone receptors has been slow. Thus, the identity of the involved G protein(s) has not been established. In many target tissues receptor-regulated phosphoinositide signaling is not sensitive to pertussis or cholera toxin (see Harden, 1989; Martin, 1989), which apparently rules out direct involvement of all of the G proteins purified to date which do serve as substrates for one or the other of these toxins. Perhaps a G protein(s) similar to Gz, which has been cloned from a human retinal cDNA library and does not possess the concensus sequence for ADP-ribosylation by bacterial toxins (Fong et al., 1988), is directly involved in receptor-PLC coupling. Alternatively, an apparently large group of 20,000–30,000 kD GTP-binding proteins of unknown function recently has been identified (Evans et al., 1986; Kikuchi et al., 1988). Although there is no precedence for this type of protein serving as a trans-plasma membrane signaling entity, this remains a possibility.

Although the work discussed above has revealed a multiplicity of PLC isozymes, it is unknown which of these, if any, are regulated by receptor-activated G proteins. Questions over the preponderance of soluble PLC activity are relevant, and indeed, data have been reported indicating that PLC activities found in soluble fractions obtained from platelets (Baldasarre and Fisher, 1986; Deckmyn et al., 1986) and thymocytes (Wang et al., 1987) are subject to activation by guanine nucleotides. Further evidence for the physical and functional interaction of soluble forms of PLC from these tissues with a soluble GTP-binding protein has been presented (Baldasarre et al., 1988; Wang et al., 1988). However, the relevance of these observations to guanine nucleotide–dependent regulation of PLC by cell surface receptors remains to be established. Advances in purifying a receptor/guanine nucleotide–regulated PLC from turkey erythrocytes are discussed in detail below.

Growth Factor Receptors and PLC

One potential form of regulation of PLC that recently has received considerable attention involves growth factor receptors. It has been known for several years that activation of platelet-derived growth factor (PDGF) (Berridge et al., 1984) and epidermal growth factor (EGF) (Hepler et al., 1987; Pike and Eakes, 1987) receptors increase phosphoinositide hydrolysis in a variety of target tissues. There was reason to believe that these receptors would not regulate PLC in a manner analogous to classical (e.g., muscarinic, angiotensin, α_1-adrenergic) hormone receptors in that the growth factor receptors show fundamental differences in structure and function (Carpenter, 1987); they express intrinsic tyrosine-specific protein kinase activity and share little or no amino acid sequence homology or predicted structure with the family of hormone and neurotransmitter receptors known to couple to G proteins (and in several cases, PLC). Wahl et al. (1988) have recently reported that anti-phosphotyrosine antibody can be used to recover PLC activity from soluble, but not particulate, extracts from EGF-treated A431 epidermoid carcinoma cells. The phosphorylated PLC is apparently the 148-kD PLC-γ in that antibodies against this, but not other, PLC isozymes precipitates tyrosine-phosphorylated protein from EGF-treated, [^{32}P]orthophosphate-labeled cells (Wahl et al., 1989; Meisenhelder et al., 1989). Similar results have recently been reported for EGF receptors in WB rat liver cells (Huckle, W. R., J. R. Hepler, S. G. Rhee, T. K. Harden, and H. S. Earp, manuscript submitted for publication) and for PDGF receptors in NIH 3T3 cells (Meisenhelder et al., 1989). EGF receptors will also tyrosine-phosphorylate purified PLC-γ, and tryptic digestion of the enzyme phosphorylated in vitro reveals a pattern of phosphorylated peptides very similar to that obtained with the enzyme phosphorylated in response to EGF in intact A431 cells (Meisenhelder et al., 1989; Nishibe et al., 1989). Similar results have been obtained with PDGF receptor-stimulated phosphorylation of PLC-γ in vitro and in intact NIH 3T3 cells (Meisenhelder et al., 1989).

The physiological significance of tyrosine phosphorylation of PLC-γ is not yet known. There are no data available that demonstrate a phosphorylation-dependent increase in catalytic activity of the enzyme, and regulation, rather, may take the form of a change in the subcellular distribution of the enzyme or a modification in interaction with some regulatory protein(s). And here, the homology of predicted structure with the src and crk oncogene products may have significance (see Rhee et al., 1989), as may the observation that phosphatidylinositol kinase activity has been found to associate with the receptor for PDGF (Whitman and Cantley, 1988; Auger et al., 1989). Along these lines, it should be pointed out that PLC-γ antibody specifically precipitates the EGF receptor along with tyrosine-phosphorylated PLC-γ in EGF-treated A431 (Wahl et al., 1989) or WB (Huckle, W. R., J. R. Hepler, S. G. Rhee, T. K. Harden, and H. S. Earp, manuscript submitted for publication) cells, and that at least three other yet-to-be-identified proteins coprecipitate in what potentially could be an EGF receptor-PLC-regulatory protein(s) complex. Other proteins also appear to be associated with PDGF receptor-regulated PLC-γ of NIH 3T3 cells (Meisenhelder et al., 1989). Whether growth factor receptors and G protein–linked receptors activate the same, different, or overlapping PLC isozymes is not known. Irrespective of the answer to this issue, data available to date are consistent with the idea that enzyme activation by the two types of receptors occurs by two fundamentally different mechanisms.

Turkey Erythrocytes as a Model for Receptor- and Guanine Nucleotide–regulated PLC

General Properties of the Guanine Nucleotide–dependent P_{2Y}-Receptor-stimulated PLC

The relatively small and short-lived response of [^{3}H]inositol-labeled mammalian membranes to guanine nucleotides or guanine nucleotides plus hormones, as well as difficulties encountered in observing hormonal regulation of PLC when labeled PtdIns(4,5)P$_2$ is presented as an exogenous substrate in lipid vesicles, led us several years ago to seek a model system that might be more amenable to the study of guanine nucleotide–dependent regulation of PLC by hormone receptors. Turkey erythrocyte membranes have proven highly useful for this purpose.

The turkey erythrocyte is a simple cell that in contradistinction to its mammalian counterparts possesses phosphatidylinositol synthase activity (Harden et al., 1987). Therefore, fresh cells can be incubated overnight with [^{3}H]inositol with resultant production of [^{3}H]-labeled phosphoinositides of high specific activity. Membranes prepared from these cells respond with marked increases in inositol phosphate production in response to the stable analogue of GTP, GTPγS, and to AlF$_4$. This membrane PLC is essentially unresponsive to Ca^{++} alone, although Ca^{++} concentrations in the 0.1–1.0-μM range enhance the response to guanine nucleotides (Harden et al., 1988).

The GTPγS-activated enzyme principally hydrolyzes PtdIns(4,5)P_2 as substrate, with a small amount of breakdown in PtdIns-4P also occurring; no hydrolysis of phosphatidylinositol is observed (Harden et al., 1988). Thus, Ins(1,4,5)P_3 is the initial inositol phosphate product, although due to the presence of Ins(1,4,5)P_3 3-kinase and 5-phosphomonoesterase activities in the membrane preparation, considerable amounts of Ins(1,3,4,5)P_4, Ins(1,3,4)P_3, and InsP_2 are formed. Little or no cyclic InsP_3 is produced under these assay conditions.

Inositol phosphate responses of intact turkey erythrocytes to a broad range of potential receptor agonists were screened with the goal of identification of a cell surface receptor capable of activating PLC. Although analysis of intact cell responses of turkey erythrocytes is difficult because of the large amount of [^{3}H]inositol labeling of inositol phosphates involved as precursors in the pathway to synthesis of InsP_5 and InsP_6, small but reproducible elevation of inositol phosphate levels could be observed in response to ATP, ADP, and analogues of ATP and ADP (Berrie et al., 1989). The pharmacological profile of these responses coincides with that expected of a P_{2Y}-purinergic receptor. The activity of this receptor can be more readily observed in membranes prepared from [^{3}H]inositol-labeled cells. In the absence of guanine nucleotides there is no effect of P_{2Y}-purinergic receptor agonists on membrane PLC activity (Boyer et al., 1989a). In the presence of GTP or a stable analogue of GTP, adenine nucleotides stimulate inositol phosphate formation with an order of potency of 2MeSATP > ADPβS > ATPγS > ATP > AppNHp = ADP > Ap(CH$_2$)pp > App(CH$_2$)p. AMP, adenosine, phenylisopropyladenosine, and 5'-dideoxyadenosine have no effect.

Kinetics of Activation of Turkey Erythrocyte PLC

The properties of regulation of PLC by guanine nucleotides and the P_{2Y}-purinergic receptor are remarkably similar to those studied previously in detail for the receptor-regulated adenylate cyclase and rhodopsin/transducin/cyclic GMP phosphodiesterase signaling systems. Thus, the concentration effect curve for GTPγS is markedly shifted to the left in a concentration-dependent and saturable manner by P_{2Y}-purinergic receptor agonists (Boyer et al., 1989a). Addition of GTPγS alone to [^{3}H]inositol-labeled membranes results in a marked increase in inositol phosphate production, although there is a considerable lag period before the attainment of maximal steady-state formation of product. P_{2Y}-purinergic receptor agonists markedly increase the rate of activation of the enzyme by stable GTP analogues, with the effects of agonists again concentration dependent, saturable, and consistent with the occurrence of a P_{2Y}-purinergic receptor-mediated event.

By using GDPβS to block further activation of the enzyme, the turn-off reaction of the GTP-, GppNHp-, and GTPγS-preactivated enzyme can be examined. Whereas high concentrations of GDPβS virtually instantaneously block GTP + agonist-stimulated enzyme activity, reversal of the GppNHp- and the GTPγS-preactivated enzyme is much slower (Boyer et al., 1989a). These differences in turn-off rates are commensurate with the differences in maximal effect of the three nucleotides on PLC activity in the presence of ADPβS.

Briefly then, a kinetic analysis of the receptor- and guanine nucleotide–regulated PLC suggests a general model for a catalytic cycle similar to that known in much greater detail for receptor- and G protein-regulated adenylate cyclase. Thus, the data are consistent with the idea that P_{2Y}-purinergic receptor agonists promote exchange of GTP and GTP analogues for GDP on the G protein involved in activation of PLC. The

GTP-liganded G protein (or more likely an activated α subunit) then interacts with the PLC catalyst to form the active enzymic species. The life time of this complex is likely governed at least in part by a GTPase activity of the G protein α subunit that returns the active species to the ground state.

Identification of the Components of the P_{2Y}-Receptor-Regulated PLC

Based on the marked and long-lived response of turkey erythrocyte PLC to P_{2Y}-purinergic receptor agonists and guanine nucleotides, as well as the advantages that such a homogeneous cell has to offer for protein purification, we have tried to identify and purify each of the protein components of this signaling system.

The P_{2Y}-receptor. A reversibly binding radioligand, [^{35}S]ADPβS, has been synthesized, and has proven very useful for identification of the turkey erythrocyte P_{2Y}-purinergic receptor in both plasma membranes (Cooper et al., 1989) and in solution after it has been solubilized with a nonionic detergent (Jeffs, R., C. L. Cooper, and T. K. Harden, manuscript submitted for publication). Preliminary data also suggest that 3'-O-(4-benzoyl)benzoyl ATP can be used as a photoaffinity agonist to irreversibly bind to the receptor and activate it in a guanine nucleotide–dependent manner (Boyer and Harden, 1989). The [^{32}P]-labeled counterpart of this affinity label can be used as a radioligand to covalently label a 50,000–55,000-kD protein that expresses the pharmacological properties of a P_{2Y}-purinergic receptor (Boyer, J. L., and T. K. Harden, manuscript submitted for publication). Only initial attempts have been made at purification of the receptor protein using these probes.

The PLC-associated G protein. Little progress has been made in identifying the involved G protein. As was discussed above there are a few hints available concerning the nature of this protein, although based on the similarity of activation/deactivation kinetics it would be a surprise if it did not share considerable functional/structural homology with the members of the G protein family that have been extensively characterized to date. In the case of the signaling system of turkey erythrocytes, a cholera or pertussis toxin substrate does not seem to be involved based on the complete lack of sensitivity of the response to inactivation by toxins (Waldo, G. L., and T. K. Harden, unpublished observations). Immunoblotting with COOH-terminal peptide antibodies indicates that G_{i3} is present in turkey erythrocyte plasma membranes (Smyth, S., and T. K. Harden, unpublished observations). However, in addition to the lack of effect of pertussis toxin intoxication, a COOH-terminal antibody that theoretically may block G_{i3}-mediated signaling events has no effect on P_{2Y}-purinergic receptor/ guanine nucleotide–stimulated PLC of turkey erythrocyte membranes (Smyth, S., and T. K. Harden, unpublished observations). Attempts to modify membrane PLC activity by reconstitution of GTPγS-preactivated α subunits of various purified G proteins or purified GTP-binding fractions obtained from turkey erythrocytes so far have not met with success.

No unambiguous information is available as to whether it is a heterotrimeric G protein that directly couples receptors to PLC. However, indirect data obtained with turkey erythrocyte membranes are consistent with this idea. G protein βγ-subunits were purified from turkey erythrocytes, bovine brain, and human placenta, reconstituted into turkey erythrocyte acceptor membranes, and effects on adenylate cyclase and PLC activities determined. Reconstituted βγ subunits inhibit both AlF$_4$-stimulated adenylate cyclase and AlF$_4$-stimulated PLC activities over the same βγ concentration range, suggesting that the inactive (heterotrimeric?) state of the PLC-associated G protein is promoted by increases in membrane $\beta\gamma$ subunit concentration (Boyer et al., 1989c). In contrast, there is an increase in P_{2Y}-purinergic receptor agonist + GTP-stimulated PLC activity after $\beta\gamma$ subunit reconstitution. A number of possibilities for such a result have been considered. For example, an increase in free $\beta\gamma$ subunit could favor formation of a heterotrimeric form of the G protein necessary for receptor-promoted exchange of GTP for GDP. Alternatively, the activity of an inhibitory G protein could be reduced by $\beta\gamma$ subunit reconstitution. Whatever the explanation for these data, they are consistent with the idea that $\beta\gamma$ subunits play a role in the PLC activation/deactivation cycle and are at least consistent with the idea that a heterotrimeric G protein is in some way involved in regulation of PLC. Moriarty et al. (1989) have reported recently that $\beta\gamma$ subunits purified from human erythrocytes and bovine brain, when injected into *Xenopus* oocytes, inhibited a muscarinic receptor-activated Cl^- current, which apparently results from PLC-catalyzed production of $Ins(1,4,5)P_3$ and Ca^{++} mobilization. These results are also consistent with the turkey erythrocyte data, which suggests the involvement of a G protein heterotrimer. Whether this protein directly couples receptors to PLC in a manner analogous to the role of G_s in coupling β-adrenergic receptors to adenylate cyclase remains to be proven.

The G protein-activated PLC. In the face of difficulties in establishing successful means of identification of the PLC-associated G protein of turkey erythrocytes, we have adopted the strategy of purification of the receptor/G protein–regulated PLC with the eventual goal of using this protein in an assay to identify its G protein cohort. Using [^{3}H]PtdIns(4,5)P_2 as a substrate in mixed phospholipid and cholate-containing micelles and a Ca^{++} concentration of 100 μM, ~83% of the total PLC activity of turkey erythrocytes can be recovered in a cytosolic fraction when intact cells are disrupted by nitrogen cavitation followed by differential centrifugation (Morris et al., 1990a). This activity has been purified 48,500-fold with good yield using ammonium sulfate precipitation followed by chromatography over Q-Sepharose, hydroxylapatite, heparin-Sepharose, and Sephacryl S-300, and a final purification step using FPLC/Mono Q. The specific activity of the purified enzyme is ~10 μmol/min per mg protein using PtdIns(4,5)P_2 as the substrate (Morris et al., 1990a). Although the turkey erythrocyte PLC is of a similar apparent size to bovine brain PLC-β and -γ, it does not react with antibody to the two brain enzymes (Waldo, G. L., and T. K. Harden, unpublished data).

A series of experiments has been performed to investigate the capacity of this purified isozyme of PLC to be regulated by the G protein–linked P_{2Y}-purinergic receptor when reconstituted with [^{3}H]inositol-labeled turkey erythrocyte ghosts. By labeling intact turkey erythrocytes to high specific radioactivity with [^{3}H]inositol and diluting membranes under conditions where they apparently lose their endogenous PLC activity, highly labeled acceptor membranes can be produced in which there is little or no measurable accumulation of inositol phosphate in response to a P_{2Y}-purinergic receptor agonist + GTPγS. Under these conditions reconstitution of as little as 1 ng of purified PLC to 5 μg of acceptor membrane results in membranes that are now markedly responsive to stimulation by P_{2Y}-purinergic receptors and guanine nucleotides (Morris et al., 1990b). With 10 μg of acceptor protein, half-maximal restoration of agonist- and guanine nucleotide–stimulated inositol lipid hydrolysis was observed with ~5 ng of purified turkey erythrocyte PLC and this effect was maximal with 100 ng. Receptor stimulation of the reconstituted enzyme shows the same absolute dependence on guanine nucleotide as does activation of the endogenous enzyme. Reconstitution of increasing amounts of enzyme resulted in coincidental increases in maximal PLC activity observed in the presence of agonist + GTPγS, with no change occurring in the rate of activation of the enzyme observed in the presence of agonist + nonhydrolyzable GTP analogue. Furthermore, the reconstituted enzyme showed inactivation kinetics in the presence of GDPβS that were identical to those for the endogenous enzyme. As with the endogenous PLC, the reconstituted enzyme was insensitive to activation by Ca^{++}, although the presence of Ca^{++} at physiological concentrations was necessary for maximal receptor and guanine nucleotide–mediated activation. The $K_{0.5}$ values for activation by agonists, guanine nucleotides, and AlF_4 were the same for the reconstituted enzyme as for the endogenous enzyme.

Although the turkey erythrocyte PLC was initially purified using an assay with exogenous [^{3}H]PtdIns(4,5)P_2 in the presence of cholate, Ca^{++}, and phospholipids, it was also important to show that reconstitution of receptor and guanine nucleotide response could be used as a means to purify the enzyme, and that this approach resulted in purification of the same enzyme as that identified with exogenous substrate assay. As such, the reconstitution approach described above was used to follow enzyme activity during purification. Conditions for reliable quantitation of reconstituted enzyme activity have not been established for the initial ammonium sulfate precipitation and Q-Sepharose steps of purification. However, at all subsequent stages of the purification, the chromatographic behavior of receptor- and guanine nucleotide–regulated PLC determined using the reconstitution assay was identical to that of PLC activity determined using exogenous [^{3}H]PtdIns(4,5)P_2 as substrate (Morris et al., 1990b).

Taken together the data obtained with PLC reconstitution suggest but do not prove that the 150-kD enzyme we have purified from turkey erythrocytes is the same enzyme as that under regulation by receptors and a putative G protein in the turkey erythrocyte plasma membrane. However, many questions remain. Is this the only PLC activity of turkey erythrocytes (it at least appears to be the predominant activity)? What is its distribution between plasma membrane and cytosolic compartments in intact cells and is there physiological significance in this distribution? Is the turkey erythrocyte enzyme capable of regulation by receptors and G proteins in mammalian plasma membranes? What is its relationship to other purified and/or cloned PLC isozymes? If it is a structurally unique protein, perhaps its G protein–regulated counterpart in mammalian tissues has yet to be purified, and structural information and molecular cloning of the turkey erythrocyte enzyme could be instrumental in identifying the PLC or class of PLC isozymes that are under regulation by guanine nucleotides and hormone receptors.

References

Auger, K. R., L. A. Serunian, S. P. Soltoff, P. Libby, and L. C. Cantley. 1989. PDGF-dependent tyrosine phosphorylation stimulates production of novel polyphosphoinositides in intact cells. *Cell.* 57:167–175.

Baldasarre, J. J., and G. J. Fisher. 1986. Regulation of membrane-associated and cytosolic phospholipase C activities in human platelets by guanosine triphosphate. *Journal of Biological Chemistry.* 261:11942–11944.

Baldasarre, J. J., P. A. Henderson, and G. J. Fisher. 1989. Isolation and characterization of one soluble and two membrane-associated forms of phosphoinositide-specific phospholipase C from human platelets. *Biochemistry.* 28:6010–6016.

Baldasarre, J. J., M. A. Knipp, P. A. Henderson, and G. J. Fisher. 1988. GTPγS-stimulated hydrolysis of phosphatidylinositol-4,5 bisphosphate by soluble phospholipase C from human platelets requires soluble GTP-binding protein. *Biochemical and Biophysical Research Communications.* 154:351–357.

Banno, Y., Y. Yada, and Y. Nozawa. 1988. Purification and characterization of membrane-bound phospholipase C specific for phosphoinositides from human platelets. *Journal of Biological Chemistry.* 263:11459–11464.

Bennett, C. F., J. M. Balcarek, A. Varrichio, and S. T. Crooke. 1988. Molecular cloning and complete amino-acid sequence of form-I phosphoinositide-specific phospholipase C. *Nature* 334:268–270.

Bennett, C. F., and S. T. Crooke. 1987. Purification and characterization of phosphoinositide-specific phospholipase C from guinea pig uterus: phosphorylation by protein kinase C in vivo. *Journal of Biological Chemistry.* 262:13789–13797.

Berridge, M. J. 1987. Inositol trisphosphate and diacylglycerol: two interacting second messengers. *Annual Review of Biochemistry.* 56:159–193.

Berridge, M. J., J. R. Heslop, R. F. Irvine, and K. D. Brown. 1984. Inositol trisphosphate formation and calcium mobilization in Swiss 3T3 cells in response to platelet-derived growth factor. *Biochemical Journal.* 222:195–201.

Berrie, C. P., P. T. Hawkins, L. R. Stephens, T. K. Harden, and C. P. Downes. 1989. Phosphatidylinositol 4,5-bisphosphate hydrolysis in turkey erythrocytes is regulated by P_{2Y}-purinoceptors. *Molecular Pharmacology.* 35:526–532.

Bloomquist, B. T., R. D. Shortridge, S. Schneuwly, M. Perdew, C. Montell, H. Steller, G. Rubin, and W. L. Pak. 1988. Isolation of a putative phospholipase C gene of *Drosophlia,* norpA, and its role in phototransduction. *Cell.* 54:723–733.

Boyer, J. L., C. P. Downes, and T. K. Harden. 1989a. Kinetics of activation of phospholipase C by P_{2Y}-purinergic receptor agonists and guanine nucleotides. *Journal of Biological Chemistry.* 264:884–890.

Boyer, J. L., J. R. Hepler, and T. K. Harden. 1989b. Hormone and growth factor receptor-mediated regulation of phospholipase C activity. *Trends in Pharmacological Sciences.* 10:360–364.

Boyer, J. L., G. L. Waldo, T. Evans, J. K. Northup, C. P. Downes, and T. K. Harden. 1989c. Modification of the AlF4—- and receptor-stimulated phospholipase C activity by G-protein βγ-subunits. *Journal of Biological Chemistry.* 264:13917–13922.

Boyer, J. L., and T. K. Harden. 1989. Irreversible activation of phospholipase C-coupled P_{2Y}-purinergic receptors by 3'-0-(4-benzoyl)benzoyl ATP. *Molecular Pharmacology.* In press.

Carpenter, G. 1987. Receptors for epidermal growth factor and other polypeptide mitogens. *Annual Review of Biochemistry.* 56:881–914.

Cooper, C. L., A. J. Morris, and T. K. Harden. 1989. Guanine nucleotide-sensitive interaction of a radiolabeled agonist with a phospholipase C-linked P_{2Y}-purinergic receptor. *Journal of Biological Chemistry.* 264:6202–6206.

Cockcroft, S., and B. D. Gomperts. 1985. Role of guanine nucleotide binding protein in the activation of polyphosphoinositide phosphodiesterase. *Nature.* 314:534–535.

Crooke, S. T., and C. F. Bennett. 1989. Mammalian phosphoinositide-specific phospholipase C isoenzymes. *Cell Calcium.* 10:309–323.

Deckmyn, H., S. M. Tu, and P. W. Majerus. 1986. Guanine nucleotides stimulate soluble phosphoinositide-specific phospholipase C in the absence of membranes. *Journal of Biological Chemistry.* 261:16553–16558.

Downes, C. P., and R. H. Michell. 1985. Inositol phospholipid breakdown as a receptor controlled generator of second messengers. *In* Molecular Mechanisms of Transmembrane Signalling. P. Cohen and M. D. Houslay, editors. Elsevier, The Netherlands. 3–56.

Evans, T., M. L. Brown, E. D. Fraser, and J. K. Northup. 1986. Purification of the major GTP-binding proteins from human placental membranes. *Journal of Biological Chemistry.* 261:7052–7059.

Fong, H. K. W., K. K. Yoshimoto, P. Eversole-Cire, and M. I. Simon. 1988. Identification of GTP-binding protein α-subunit that lacks an apparent ADP-ribosylation site for pertussis toxin. *Proceedings of the National Academy of Sciences.* 85:3066–3070.

Fukui, T., R. J. Lutz, and J. M. Lowenstein. 1988. Purification of a phospholipase C from rat liver cytosol that acts on phosphatidylinositol 4,5-bisphosphate and phosphatidylinositol 4-phosphate. *Journal of Biological Chemistry.* 263:17730–17737.

Hakata, H., J.-L. Kambayashi, and G. Kosaki. 1982. Purification and characterization of phosphatidylinositol-specific phospholipase C from human platelets. *Journal of Biochemistry.* 92:929–935.

Harden, T. K. 1989. The role of guanine nucleotide regulatory proteins in receptor-selective direction of second messenger signalling. *In* Inositol Lipids in Cell Signalling. R. H. Michell, A. H. Drummond, and C. P. Downes, editors. Academic Press. London. 113–134.

Harden, T. K., P. T. Hawkins, L. Stephens, J. L. Boyer, and C. P. Downes. 1988. Phosphoinositide hydrolysis by guanosine 5′-O-(3-thiotriphosphate)–activated phospholipase C of turkey erythrocyte membranes. *Biochemical Journal.* 252:583–593.

Harden, T. K., L. Stephens, P. T. Hawkins, and C. P. Downes. 1987. Turkey erythrocyte membranes as a model for regulation of phospholipase C by guanine nucleotides. *Journal of Biological Chemistry.* 262:9057–9061.

Hepler, J. R., N. Nakahata, T. W. Lovenberg, J. DiGuiseppi, B. Herman, H. S. Earp, and T. K. Harden. 1987. Epidermal growth factor stimulates the rapid accumulation of inositol (1,4,5) trisphosphate and a rise in cytosolic calcium mobilized from intracellular stores in A431 cells. *Journal of Biological Chemistry.* 262:2951–2956.

Herrero, C., M. E. Cornet, C. Lopez, P. G. Barreno, A. M. Municico, and J. Moscat. 1988. Ca^{2+}-induced changes in the secondary structure of a 60 kDa phosphoinositide-specific phospholipase C from bovine brain cytosol. *Biochemical Journal.* 255:807–812.

Hirasawa, K., R. F. Irvine, and R. M. Dawson. 1981. The hydrolysis of phosphatidylinositol monolayers at an air/water interface by the calcium ion dependent phosphatidylinositol phosphodiesterase of pig brain. *Biochemical Journal.* 193:607–614.

Hofmann, S. L., and P. W. Majerus. 1982. Modulation of phosphatidylinositol-specific phospholipase C activity by phospholipid interactions, diglycerides, and calcium ions. *Journal of Biological Chemistry.* 257:14359–14364.

Homma, Y., J. Imaki, O. Nakanishi, and T. Takenawa. 1988. Isolation and characterization of two different forms of inositol phospholipid-specific phospholipase C from rat brain. *Journal of Biological Chemistry.* 263:6592–6598.

Irvine, R. F., N. Hemington, and R. M. C. Dawson. 1979a. The calcium-dependent phosphatidylinositol phosphodiesterase of rat brain: Mechanisms of suppression and stimulation. *European Journal of Biochemistry.* 99:525–530.

Irvine, R. F., A. J. Letcher, and R. M. C. Dawson. 1979b. Fatty acid stimulation of membrane phosphatidylinositol hydrolysis by brain phosphatidylinositol phosphodiesterase. *Biochemical Journal.* 178:497–500.

Katan, M., R. W. Kriz, N. Totty, R. Philip, E. Meldrum, R. A. Aldape, J. L. Knopf, and P. J. Parker. 1988. Determination of the primary structure of PLC-154 demonstrates diversity of phosphoinositide-specific phospholipase C activities. *Cell.* 54:171–177.

Kikuchi, A., T. Yamashita, M. Kawata, K. Yamamoto, K. Ikeda, T. Tanimoto, and Y. Takai. 1988. Purification and characterization of a novel GTP-binding protein with a molecular weight of 24,000 from bovine brain membranes. *Journal of Biological Chemistry.* 263:2897–2904.

Lee, K. Y., S. H. Ryu, P. G. Suh, W. C. Choi, and S. G. Rhee. 1987. Phospholipase C associated with particulate fractions of bovine brain. *Proceedings of the National Academy of Sciences.* 84:5540–5554.

Linden, J., and T. M. Delahunty. 1989. Receptors that inhibit phosphoinositide breakdown. *Trends in Pharmacological Sciences.* 10:114–120.

Litosch, I., C. Wallis, J. N. Fain. 1985. 5-Hydroxytryptamine stimulates inositol phosphate production in a cell-free system from blowfly salivary glands. *Journal of Biological Chemistry.* 260:5464–5471.

Martin, T. F. J. 1989. Lipid hydrolysis by phosphoinositidase C: enzymology and regulation by receptors and guanine nucleotides. *In* Inositol Lipids in Cell Signalling. R. H. Michell, A. H. Drummond, and C. P. Downes, editors. Academic Press, London. 81–107.

Meisenhelder, J., P.-G. Suh, S. G. Rhee, and T. Hunter. 1989. Phospholipase C-γ is a substrate for the PDGF receptor protein-tyrosine kinases in vivo and in vitro. *Cell.* 56:1009–1122.

Meldrum, E., M. Katan, and P. Parker. 1989. A novel inositol-phospholipid-specific phospholipase C. *European Journal of Biochemistry.* 182:673–677.

Michell, R. H. 1975. Inositol phospholipids and cell surface receptor function. *Biochimica et Biophysica Acta.* 415:81–147.

Moriarty, T. M., B. Gillo, D. J. Carty, R. T. Premont, E. M. Landau, and R. Iyengar, 1988. $\beta\gamma$ subunits of GTP-binding proteins inhibit muscarinic receptor stimulation of phospholipase C. *Biochemistry.* 85:8865–8869.

Morris, A. J., G. L. Waldo, C. P. Downes, and T. K. Harden. 1990a. A receptor and G-protein-regulated polyphosphoinositide specific phospholipase C. 1. Purification and properties. *Journal of Biological Chemistry.* In press.

Morris, A. J., G. L. Waldo, C. P. Downes, and T. K. Harden, 1990b. A receptor and G-protein regulated polyphosphoinositide-specific phospholipase C from turkey erythrocytes. 2. P_{2Y}-purinergic receptor and G-protein-mediated regulation of the purified enzyme. *Journal of Biological Chemistry.* In press.

Nishibe, S., M. I. Wahl, S. G. Rhee, and G. Carpenter. 1989. Tyrosine phosphorylation of phospholipase C-II in vitro by the epidermal growth factor receptor. *Journal of Biological Chemistry.* 264:10335–10338.

Ohta, S., A. Matsui, Y. Nazawa, and Y. Kagawa. 1989. Complete cDNA encoding a putative phospholipase C from transformed human lymphocytes. *Federation of European Biochemical Societies Letters.* 242:31–35.

Pike, L. J., and A. T. Eakes. 1987. Purification of a phosphoinositide-specific phospholipase C from bovine brain. Epidermal growth factor stimulates the production of phosphatidylinositol monophosphate and the breakdown of polyphosphoinositides in A431 cells. *Journal of Biological Chemistry.* 262:1644–1651.

Rebecchi, M. J., and O. M. Rosen. 1987. Purification of a phosphoinositide-specific phospholipase C from bovine brain. *Journal of Biological Chemistry.* 262:12526–12532.

Rhee, S. G., P. G. Suh, S.-H. Ryu, and S. Y. Lee. 1989. Studies of inositol phospholipid-specific phospholipase C. *Science.* 244:546–550.

Ryu, S. H., K. S. Cho, K.-Y. Lee, P.-G. Suh, and S. G. Rhee. 1987. Purification and characterization of two immunologically distinct phosphoinositide-specific phospholipase C from bovine brain. *Journal of Biological Chemistry.* 262:12511–12518.

Stahl, M. L., C. R. Ferenz, K. L. Kelleher, R. W. Kriz, and J. L. Knopf. 1988. Sequence similarity of phospholipase C with the noncatalytic region of src. *Nature.* 332:269–272.

Suh, P.-G., S. H. Ryu, W. C. Choi, K.-Y. Lee, and S. G. Rhee. 1988a. Monoclonal antibodies to three phospholipase C isozymes from bovine brain. *Journal of Biological Chemistry.* 263:144797–145804.

Suh, P.-G., S. H. Ryu, K. H. Moon, H. W. Suh, and S. G. Rhee. 1988b. Inositol phospholipid-specific phospholipase C; complete CDNA and protein sequences and sequence homology to tyrosine kinase-related oncogene products. *Proceedings of the National Academy of Sciences.* 85:5419–5423.

Suh, P.-G., S. H. Ryu, K. H. Moon, H. W. Suh, and S. G. Rhee. 1988c. Cloning and sequence of multiple forms of phospholipase C. *Cell.* 54:161–169.

Wahl, M. I., T. O. Daniel, and G. Carpenter. 1988. Antiphosphotyrosine recovery of phospholipase C activity after EFG treatment of A-431 cells. *Science.* 241:986–970.

Wahl, M. I., S. Nishibe, P.-G. Suh, S. G. Rhee, and G. Carpenter. 1989. Epidermal growth factor stimulates tyrosine phosphorylation of phospholipase C-II independently of receptor internalization and extracellular calcium. *Proceedings of the National Academy of Sciences.* 86:1569–1572.

Wang, P., S. Toyoshima, and T. Osawa. 1987. Physical and functional association of cytosolic inositol-phospholipid-specific phospholipase C of calf thymocytes with a GTP-binding protein. *Journal of Biochemistry.* 102:1275–1287.

Wang, P., S. Toyoshima, and T. Osawa, 1988. Properties of a novel GTP-binding protein which is associated with soluble phosphoinositide specific phospholipase C. *Journal of Biochemistry.* 103:137–142.

Whitman, M., and L. Cantley. 1988. Phosphoinositide metabolism and the control of proliferation. *Biochimica et Biophysica Acta.* 948:327–344.

Chapter 6

Structure/Function Relationships of Ras and
Guanosine Triphosphatase–activating Protein

Jackson B. Gibbs, Michael D. Schaber, Victor M. Garsky,
Ursula S. Vogel, Edward M. Scolnick, Richard A. F. Dixon,
and Mark S. Marshall

*Departments of Molecular Biology and Medicinal Chemistry, Merck,
Sharp and Dohme Research Laboratories, West Point,
Pennsylvania 19486*

Introduction

Two general classes of GTP-binding regulatory proteins (G proteins) are involved with a diversity of signal transduction processes. The 40-kD G protein class is characterized by proteins having heterotrimeric complexes composed of 40-kD (α, binds GTP), 35 kD (β), and 8 kD (γ) subunits (Gilman, 1987). These G proteins typically couple receptors to enzymes and ion channels and are associated with the transduction of extracellular hormone signals into changes of intracellular metabolites such as cAMP, cGMP, or phosphoinositides. The 20-kD G protein class is characterized by monomeric proteins that are associated with a variety of functions such as secretion, microtubule and microfilament organization, and other processes central to normal cellular homeostasis, but are not yet clearly coupled to specific receptors and enzymes (Chardin et al., 1989; Gibbs and Marshall, 1989). Regulation of cellular processes by G proteins is conserved in evolution as evidenced by the diversity of G proteins in lower eucaryotic organisms such as the yeast and slime mold (Gibbs and Marshall, 1989). In the yeast *Saccharomyces cerevisiae,* the essential functions of the different 20-kD G proteins are not complemented by other members of the 20-kD G protein family, indicating that there is high specificity due to either subcellular localization or specific binding interactions with target molecules.

At least 13 different 40-kD Gα subunits and 23 different 20-kD G proteins have been identified (Chardin et al., 1989; Didsbury et al., 1989; Strathmann et al., 1989). The function of proteins in both G protein classes is dependent on GTP binding. Upon hydrolysis of the bound GTP to GDP, the G protein ceases to stimulate the relevant metabolic process under its regulation. The 189 amino acid product of the *ras* gene, Ras or p21, was the first G protein identified that could transform normal cells into cancer cells. Transforming *ras* alleles are identified in human cancers with frequencies of 15–90% depending on the tumor type. More recently, there have been reports implicating another 20-kD G protein, Rho, and a 40-kD G protein, G_s, in tumorigenesis (Avraham and Weinberg, 1989; Landis et al., 1989). The biochemical mechanisms that activate the *ras* protein have provided the best evidence that the 20-kD class of G proteins function by a GTP/GDP cycle in a manner analogous to that described for 40-kD G proteins. These naturally occurring mutations either impair GTP hydrolytic activity or facilitate nucleotide exchange leading to a constitutively active Ras-GTP complex.

The striking similarity in the crystal structures of bacterial elongation factor Tu (EF-Tu) G domain and Ras indicates a highly conserved motif among GTP-binding proteins not only for residues critical to nucleotide binding but also in the overall folding and tertiary structure (Pai et al., 1989; Tong et al., 1989). Since the 20-kD G proteins have diverse yet highly specific functions, a key question is what determinants are involved with G protein interaction with target proteins. Structure-function studies of Ras have identified a region having residues 32–40 in which mutations or deletions render the *ras* allele transformation-defective (Sigal et al., 1988). The mutant proteins expressed by these alleles bind GTP and localize to the plasma membrane, properties that are required for Ras function. This result led to speculation that this region might be involved with protein-protein interaction, possibly the effector molecule.

The identification of proteins with which Ras interacts is obviously central to our understanding of Ras function. In the yeast *S. cerevisiae,* Ras stimulates adenylyl cyclase and appears to be directly coupled to this enzyme (Gibbs and Marshall, 1989).

The *CDC25* and *IRA* gene products are involved with regulation of *RAS* protein activity. CDC25 acts in a positive manner to promote formation of the active RAS-GTP complex, possibly as a nucleotide exchange factor. IRA acts in a negative manner to promote formation of the biochemically inactive RAS-GDP complex either by stimulating RAS GTP hydrolytic activity or by inhibiting formation of the RAS-GTP complex. At present, it is not clear whether CDC25 or IRA bind directly to RAS. Genetic and biochemical approaches have yielded a great deal of information on the proteins associated with RAS function in *S. cerevisiae;* however, regulation of adenylyl cyclase activity by RAS appears to be specific to this species.

GAP

In higher eucaryotic organisms such as mammals and *Xenopus,* only one protein has been discovered so far to interact with Ras. The GTPase-activating protein (GAP) was discovered by Trahey and McCormick (1987) as a factor in cytosolic extracts of *Xenopus* oocytes and cultured mammalian cells that could stimulate the GTP hydrolytic activity of Ras at least 100-fold. Activated forms of Ras having impaired GTPase were not responsive to GAP, suggesting that this factor was central to Ras function. Two mechanisms have been proposed for GAP function—an upstream negative regulatory protein and the immediate downstream effector of Ras. To evaluate Ras/GAP interactions with Ras proteins unresponsive to GAP enzymatically, a kinetic competition assay was developed. The GTP complexes of both normal and oncogenic forms of Ras bind to GAP with micromolar affinity, about 100-fold more tightly than the biologically inactive GDP complexes (Vogel et al., 1988; Schaber et al., 1989). This type of binding interaction is similar to what one might predict for Ras binding to a target protein, although this is vitro biochemical information certainly is not proof of GAP as the Ras effector.

An extension of this correlation was provided by Adari et al. (1988) and Cales et al. (1988) who observed that forms of Ras having mutations in the region 32–40 that impair biological function also render Ras insensitive to GAP. To evaluate these observations in further detail, we tested the ability of these mutant proteins to bind to GAP. In addition, we compared the results with the ability of these Ras proteins to interact with *S. cerevisiae* adenylyl cyclase (Schaber et al., 1989). As shown in Table I, in general, if a mutation impairs Ras transforming activity, it also impairs the ability of Ras to bind GAP and to reconstitute yeast adenylyl cyclase activity. An apparent exception is the Thr to Ser substitution at residue 35, indicating that Ras interaction with other proteins in vitro does not perfectly mimic the analogous interaction in vivo.

GAP, as characterized by protein purification and cloning of the cDNA, is predicted to encode a 1,044 amino acid polypeptide of 116 kD and appears on SDS-PAGE as 120–125 kD (Gibbs et al., 1988; Vogel et al., 1988; Trahey et al., 1989). By gel filtration, GAP is apparently a monomeric polypeptide. The amino acid sequence of GAP as predicted by the cDNA reveals a striking similarity between GAP and a previously described motif termed Src homology 2 (SH2). The SH2 domain in GAP is duplicated at residues 178–256 and 348–426. The SH2 motif has been observed in the noncatalytic domain of nonreceptor tyrosine kinases such as Src and Fps, the *crk* oncogene product, and phosphoinositide-specific phospholipase C-γ. The SH2 homology in Crk is particularly intriguing because this protein is essentially a *gag* fusion with SH2 and another motif termed SH3. The SH3 region is not essential for

TABLE I
The Effect of Ras Mutations on Ras Biological and Biochemical Functions

Ras effector region	Foci	GAP binding	Yeast ADC reconstitution
32 40 | | Y D P T I E D S Y	+	+	+
. N	−	−	−
. . . A	−	−	−
. . . S	−	+	+
. . . . . . N . .	−	−	−
. . . . . . A . .	−	−	−

Ras proteins having mutations in the region 32–40 were assayed for the ability to transform NIH3T3 cells (foci), to interact with GAP, or to reconstitute adenylyl cyclase activity in membranes isolated from the *S. cerevisiae* strain F1D (*ras1 ras2 CRI4*). Results are summarized from Sigal et al., 1988 and Schaber et al., 1989.

Crk transforming activity whereas the SH2 region is critical (Mayer et al., 1988). If one draws an analogy between Crk and GAP, the SH2 regions in GAP might have an effector function associated with cellular proliferation.

A deletion analysis of the GAP cDNA has identified a COOH-terminal 343 amino acid domain that contains all of the determinants for Ras binding and stimulation of Ras GTPase activity (Marshall et al., 1989). The minimal fragment encodes [702-1044]GAP. Upon expression and purification from *Escherichia coli*, this truncated GAP polypeptide had the same affinity for Ras and could stimulate Ras GTPase with the same specific activity as that observed for full-length GAP purified from bovine brain. The identification of a distinct Ras-binding domain in the COOH-terminal half of GAP and the SH2 domain in the NH_2-terminal half of GAP raises a multidomain hypothesis of GAP function. In the model of GAP as a regulator, the SH2 region in vivo might influence activity of the COOH-terminal Ras-binding domain even though an interaction is not apparent in vitro. In the model of GAP as downstream of Ras, the SH2 domain might become activated upon Ras binding to GAP. The similarity of Ras binding to GAP and adenylyl cyclase, a known target of Ras, suggests that GAP is either the effector of Ras or that it can bind to the "same" site as the effector. A model of these two possibilities is shown in Fig. 1.

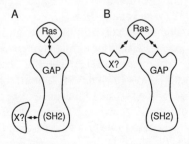

Figure 1. Models of GAP function. GAP has two distinct domains: an NH_2-terminal Src homology 2 (SH2) region and a COOH-terminal Ras-binding domain. (*A*) GAP may be the immediate target protein of Ras, possibly coupling to other proteins via the SH2 domain. (*B*) GAP may regulate Ras activity in competition with the true effector protein of Ras.

TABLE II
Peptides Inhibit Ras/GAP Interaction

Peptide		Inhibition
Ras 17–44	SALTIQLIQNHFVDEYDPTIEDSYRKQV	+
Ras 17–37	SALTIQLIQNHFVDEYDPTIE	+
Ras 17–32	SALTIQLIQNHFVDEY	+
Ras 17–26	SALTIQLIQN	−
Ras 23–37	LIQNHFVDEYDPTIE	−
Ras 31–43	EYDPTIEDSYRKQ	−

The indicated Ras peptides were assayed for inhibition of GAP-stimulated Ras GTPase activity. Results are summarized from Schaber et al., 1989. +, active at 1–4 μM; −, not active up to 250 μM.

Peptide Inhibitors of Ras/GAP Interaction

To analyse Ras interaction with GAP by another approach, Ras peptides spanning the region 32–40 were tested (Schaber et al., 1989), and the results are summarized in Table II. Ras peptides 17–44, 17–37, and 17–32 inhibited the Ras/GAP GTPase assay with affinities of 1–4 μM. A smaller fragment, peptide 17–26, was inactive. Surprisingly, peptides lacking residues 33–40 were active. This result suggested that another region of Ras adjacent to this 32–40 region might also be important for Ras binding to another protein. This region is highlighted on the crystal structure of Ras shown in Fig. 2; residues 26–32 are well exposed. The guanine nucleotide phosphates lie beneath this region and may influence the conformation as Ras switches between the GTP and GDP complexes. The inability of Ras peptides 23–37 and 31–43 to inhibit Ras/GAP interactions does not necessarily contradict the results of the mutagenesis because it is possible that these peptides do not adopt a critical conformation.

The possibility raised by the peptide results, that another region might be involved with Ras binding to GAP, is intriguing when one compares the sequence of Ras to other 20-kD G proteins (Fig. 3). For example, Ras transforms cells, and Rap (Rap1A/K-rev) suppresses the transformed phenotype (Kitayama et al., 1989). Furthermore, these proteins are responsive to chromatographically distinct GAP molecules (Kikuchi et al., 1989). Ras and Rap share amino acid identity in the region 32–40 but are divergent in the region 23–31. This divergent region may influence the different functions of these two proteins.

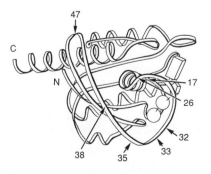

Figure 2. Structure of Ras complexed to GDP. Highlighted on the crystal structure of Ras-GDP are regions required for Ras/GAP interaction. See text for details. Adapted from Tong et al., 1989, *Science,* 245:244. Copyright 1989 by the AAAS.

GAP Biology

The in vitro biochemical analyses of Ras/GAP interactions are unable to prove the function of GAP. As a first step toward biological analysis of GAP, we have begun to test for phenotypes in two systems, *S. cerevisiae* cells and *Xenopus* oocytes. The yeast *S. cerevisiae* has a gene product, IRA, that acts as an upstream negative regulatory element of yeast RAS function. IRA also shares a striking amino acid similarity with GAP in the COOH-terminal Ras-binding domain of GAP. This homologous region most likely contains determinants essential for Ras binding. Consistent with this hypothesis, GAP peptide 888–910, which spans the homologous region, inhibits Ras/GAP interaction. To test whether mammalian GAP could substitute for IRA function, mammalian GAP was expressed in yeast strains lacking functional IRA. Both full-length GAP and the COOH-terminal domain of GAP (664–1044) suppressed the *ira* defect (Ballester et al., 1989; Tanaka et al., 1990). Thus, in yeast, GAP clearly serves an upstream negative regulatory role in RAS function. However, the NH_2-terminal half of GAP having the SH2 region is apparently not required for this GAP function. The question therefore remains what is the function of GAP in a higher eucaryotic organism.

```
              20   25   30   35   40   45   50       GAP    foci
              |    |    |    |    |    |    |
ras     TIQLIQNHFVDEYDPTIEDSYRKQVV-IDGET              A      +
R-ras   TIQFIQSYFVSDYDPTIEDSYTKICS-VDGIP              A      -
rap1    TVQFVQGIFVEKYDPTIEDSYRKQV-EVDCQQ              B      -
rho     LIVFSKDQFPEVYVPTVFENYVADI-EVDGKQ              C     (+)
ral     TVQFMYDEFVEDYEPTKADSYRKKVV-LDGEE              ?      -
YPT1    LLRFADDTYTESYISTIGVDFKIRTIELDGK               ?      -
SEC4    LVRFVEDKFNPSFITTIGIDFKIKTVDINGK               ?      -
              *    *    *    *              **
```

Figure 3. Primary structure comparison of 20-kD G proteins in the region essential to Ras effector function. Where indicated, these proteins are sensitive to chromatographically distinct GAP molecules (Gibbs et al., 1988; Garrett et al., 1989; Kikuchi et al., 1989) and can induce tumor (foci) formation (Avraham and Weinberg, 1989; Gibbs and Marshall, 1989). *, Residues highly conserved among these G proteins.

Xenopus oocytes was chosen for further biological studies because microinjection experiments are easy relative to using mammalian cells. In a model of GAP as a possible downstream effector of Ras, one needs to reconcile the intracellular localization of these two proteins. Ras must be localized in the membrane and >95% is present in the particulate fraction. GAP activity and immunoreactivity is >99% in the cytosol. Thus, one would predict that GAP function would depend on interaction with membrane-bound Ras. If GAP were unable to interact with membrane-bound Ras, one would also predict that this perturbation would block Ras-mediated cellular responses.

To test this idea, we microinjected a mutant form of yeast RAS1 that lacks the membrane-localization consensus site, is insensitive to GAP in vitro because it has a GTPase-impairing mutation, but binds tightly to GAP (Gibbs et al., 1989). The cytosol-localized RAS1 mutant was called [Leu-68]RAS1(term.) and alone did not induce germinal vesicle breakdown (GVBD). Instead, [Leu-68]RAS1(term.) was observed to inhibit the ability of mammalian [Val-12]Ras to induce GVBD in *Xenopus* oocytes by at least 80% in a dose-dependent manner. The effect of [Leu-68]RAS1(term.)

TABLE III
Microinjection of Proteins into *Xenopus* Oocytes

	Germinal vesicle breakdown	
	−	+ [Val-12]Ras
	% of [Val-12]Ras response	
Buffer	0	100
GAP	0	98
RAS1 (term.)	0	102
[Leu-68]RAS1(term.)	0	40
[Leu-68]RAS1(term.) + GAP	ND	84

[Val-12]Ras (40 μM) and yeast RAS1 proteins (30 μM), purified from *E. coli*, were equilibrated with GTP before injection. Bovine GAP (15 μM) was also purified from *E. coli*. Proteins were premixed before injection, and germinal vesicle breakdown was scored 21 h after injection. Details can be found in Gibbs et al., 1989.

at 30 μM is shown in Table III. A GAP-sensitive control, RAS1(term.), was not inhibitory.

Microinjection of purified GAP alone did not have any apparent biological effect (Table III), suggesting that GAP concentrations are not rate-limiting in *Xenopus* oocytes. However, premixing 15 μM purified mammalian GAP with 30 μM [Leu-68]RAS1(term.) could reduce the inhibitory effect of [Leu-68]RAS1(term.) on [Val-12]Ras-mediated GVBD, suggesting that GAP might be at least one intracellular target of [Leu-68]RAS1(term.). [Leu-68]RAS1(term.) also inhibits GAP activity in vivo, although the inhibition is not as striking as the inhibition of GVBD and requires higher [Leu-68]RAS1(term.) concentrations (Table IV). The poorer inhibition of GAP enzymatic activity in vivo may reflect the micromolar potency of [Leu-68]RAS1(term.) or raise the possibility that the biological target of [Leu-68]RAS1(term.) is a protein other than GAP. Antibody Y13-259, which binds to Ras

TABLE IV
Ras GTPase Activity in Living *Xenopus* Oocytes

Coinjection	Total	GTP	GDP	Conversion
	cpm			%
Buffer	1418	166	1252	88
RAS1 (term.)	1497	115	1382	92
[Leu-68]RAS1(term.)	1422	606	816	57
Ab Y13-259	1347	1347	0	0

Normal Ha Ras-[α-^{32}P]GTP (10 nM) was coinjected into *Xenopus* oocytes with the indicated yeast RAS1 protein complexed with GTP (2.5 mM) or with 1 mg/ml anti–Ras monoclonal antibody Y13-259. After 10 min at 21°C, the oocytes were lysed in buffer containing 100 μg/ml Y13-259 (see Gibbs et al., 1987, for Methods), Ras was immunoprecipitated with Protein A-Sepharose, and the bound guanine nucleotide was analyzed by chromatography on PEI-cellulose. After autoradiographic visualization, radioactive spots were scraped and quantitated, and the conversion of GTP to GDP was calculated. Results were corrected for the amount of [α-^{32}P]GDP bound to Ras before injection (200 cpm).

with at least nanomolar affinity and which inhibits Ras/GAP interaction, was completely inhibitory.

Future Directions
Studies of 20-kD G proteins obviously need to focus on the functions of these proteins and the mechanisms that regulate these processes. The characterization of the effector proteins will enable a better biochemical comparison between the fundamental activities of 20-kD and 40-kD G proteins. In studies of Ras, this work will also enable the development of inhibitors that will disrupt Ras function. Such inhibitors might have pharmacological applications in cancer chemotherapy.

References
Adari, H., D. R. Lowy, B. M. Willumsen, C. J. Der, and F. McCormick. 1988. Guanosine triphosphatase activating protein (GAP) interacts with the p21 effector binding domain. *Science.* 240:518–521.

Avraham, H., and R. A. Weinberg. 1989. Characterization and expression of the human *rhoH12* gene product. *Molecular and Cellular Biology.* 9:2058–2066.

Ballester, R., T. Michaeli, K. Ferguson, H.-P. Xu, F. McCormick, and M. Wigler. 1989. Genetic analysis of mammalian GAP expressed in yeast. *Cell.* 59:681–686.

Cales, C., J. F. Hancock, C. J. Marshall, and A. Hall. 1988. The cytoplasmic protein GAP is implicated as the target for regulation by the *ras* gene product. *Nature.* 332: 548–551.

Chardin, P., N. Touchot, A. Zahraoui, V. Pizon, I. Lerosey, B. Olofsson, and A. Tavitian. 1989. Structure of the human *ras* gene family. *In* The Gaunine Nucleotide Binding Proteins: Common Structural and Functional Properties. L. Bosch, B. Kraal, and A. Parmeggiani, editors. Plenum Publishing Co., New York. p. 153–163.

Didsbury, J., R. F. Weber, G. M. Bokoch, T. Evans, and R. Synderman. 1989. *rac,* a novel *ras*-related family of proteins that are Botulinum toxin substrates. *Journal of Biological Chemistry.* 264:16378–16382.

Garrett, M. D., A. J. Self, C. van Oers, and A. Hall. 1989. Identification of distinct cytoplasmic targets for *ras*/R-*ras* and *rho* regulatory proteins. *Journal of Biological Chemistry.* 264:10–13.

Gibbs, J. B., and M. S. Marshall. 1989. The *ras* oncogene- an important regulatory element in lower eucaryotic organisms. *Microbiological Reviews.* 53:171–185.

Gibbs, J. B., M. D. Schaber, W. J. Allard, I. S. Sigal, and E. M. Scolnick. 1988. Purification of ras GTPase activating protein from bovine brain. *Proceedings of the National Academy of Sciences.* 85:5026–5030.

Gibbs, J. B., M. D. Schaber, M. S. Marshall, E. M. Scolnick, and I., S. Sigal. 1987. Identification of guanine nucleotides bound to *ras*-encoded proteins in growing yeast cells. *Journal of Biological Chemistry.* 262:10426–10429.

Gibbs, J. B., M. D. Schaber, T. L. Schofield, and E. M. Scolnick, and I. S. Sigal. 1989. *Xenopus* oocyte germinal-vesicle breakdown induced by [Val^{12}]Ras is inhibited by a cytosol-localized Ras mutant. *Proceedings of the National Academy of Sciences.* 86:6630–6634.

Gilman, A. G., 1987. G-proteins: transducers of receptor-generated signals. *Annual Review of Biochemistry.* 56:615–650.

Kikuchi, A., T. Sasaki, S. Araki, Y. Hata, and Y. Takai. 1989. Purification and characterization from brain cytosol of two GTPase-activating proteins specific for smg p21, a GTP-binding protein having the same effector domain as c-ras p21s. *Journal of Biological Chemistry.* 264:9133–9136.

Kitayama, H., Y. Sugimoto, T. Matsuzaki, Y. Ikawa, and M. Noda. 1989. A *ras*-related gene with transformation suppressor activity. *Cell.* 56:77–84.

Landis, C. A., S. B. Masters, A. Spada, A. M. Pace, H. R. Bourne, and L. Vallar. 1989. GTPase inhibiting mutations activate the α chain of G_s and stimulate adenylyl cyclase in human pituitary tumors. *Nature.* 340:692–696.

Marshall, M. S., W. S. Hill, A. S. Ng, U. S. Vogel, M. D. Schaber, E. M. Scolnick, R. A. F. Dixon, I. S. Sigal, and J. B. Gibbs. 1989. A C-terminal domain of GAP is sufficient to stimulate *ras* p21 GTPase activity. *EMBO Journal.* 8:1105–1109.

Mayer, B. J., M. Hamaguchi, and H. Hanafusa. 1988. Characterization of p47[gag-crk], a novel oncogene product with sequence similarity to a putative modulatory domain of protein-tyrosine kinases and phospholipase C. *Cold Spring Harbor Symposia on Quantitative Biology.* 53:907–914.

Pai, E. F., W. Kabsch, U. Krengel, K. C. Holmes, J. John, and A. Wittinghofer. 1989. Structure of the guanine-nucleotide–binding domain of the Ha-*ras* oncogene product p21 in the triphosphate conformation. *Nature.* 341:209–214.

Schaber, M. D., V. M. Garsky, D. Boylan, W. S. Hill, E. M. Scolnick, M. S. Marshall, I. S. Sigal, and J. B. Gibbs. 1989. Ras interaction with the GTPase-activating protein (GAP). *Proteins: Structure, Function, and Genetics.* 6:306–315.

Sigal, I. S., M. S. Marshall, M. D. Schaber, U. S. Vogel, E. M. Scolnick, and J. B. Gibbs. 1988. Structure-function studies of the *ras* protein. *Cold Spring Harbor Symposia on Quantitative Biology.* 53:863–869.

Strathmann, M., T. M. Wilkie, and M. I. Simon. 1989. Diversity of the G-protein family: Sequences from five additional α subunits in the mouse. *Proceedings of the National Academy of Sciences.* 86:7407–7409.

Tanaka, K., M. Nakafuku, T. Satoh, M. S. Marshall, J. B. Gibbs, K. Matsumoto, Y. Kaziro, and A. Toh-e. 1990. *S. cerevisiae* genes *IRA1* and *IRA2* encode proteins that may be functionally equivalent to mammalian *ras* GTPase activating protein. *Cell.* 60:803–807.

Tong, L., M. V. Milburn, S. M. de Vos, and S.-H. Kim. 1989. Structure of *ras* protein. *Science.* 245:244.

Trahey, M., and F. McCormick. 1987. A cytoplasmic protein stimulates normal N-*ras* p21 GTPase, but does not affect oncogenic mutants. *Science.* 238:542–545.

Trahey, M., G. Wong, R. Jalenbeck, B. Rubinfeld, G. A. Martin, M. Ladner, C. M. Long, W. J. Crosier, K. Watt, K. Koths, and F. McCormick. 1989. Molecular cloning of two types of GAP complementary DNA from human placenta. *Science.* 242:1697–1700.

Vogel, U. S., R. A. F. Dixon, M. D. Schaber, R. E. Diehl, M. S. Marshall, E. M. Scolnick, I. S. Sigal, and J. B. Gibbs. 1988. Cloning of bovine GAP and its interaction with oncogenic *ras* p21. *Nature.* 335:90–93.

Chapter 7

Regulation of G Protein–coupled Receptors by Agonist-dependent Phosphorylation

Jeffrey L. Benovic, James J. Onorato, Marc G. Caron, and Robert J. Lefkowitz

Fels Institute for Cancer Research and Molecular Biology, Temple University School of Medicine, Philadelphia, Pennsylvania 19140; and the Howard Hughes Medical Institute, Duke University Medical Center, Durham, North Carolina 27710

Introduction

The β-adrenergic receptor, a 64,000 D transmembrane glycoprotein, mediates many of the actions of the catecholamines epinephrine and norepinephrine via its G protein–mediated stimulation of the enzyme adenylyl cyclase. This receptor has become the prototype for studies of the large super family of G protein–coupled receptors (Dohlman et al., 1987a). An interesting property of this system is that the functional effectiveness of the receptor in stimulating adenylyl cyclase becomes attenuated in the face of persistent agonist stimulation. This phenomenon, referred to as desensitization, appears to have a number of molecular mechanisms (Benovic et al., 1988a). In addition, such desensitization phenomena are quite widespread in nature, having been observed in systems as diverse as mating responses of yeast, chemotactic responses in slime mold, and numerous examples in higher eukaryotic organisms. The great diversity of such desensitization phenomena and the fact that they arose so early in evolution and have persisted suggests that they are of fundamental regulatory significance.

Numerous regulatory phenomena in biology are mediated, at the molecular level, by covalent modification of proteins such as enzymes, ion channels, etc. Such covalent modifications often serve to control the enzymatic or ion translocating properties of these molecules. Among the covalent modifications which regulate such functions, one of the most prevalent is phosphorylation/dephosphorylation. Studies over the last few years have demonstrated that G protein–coupled receptors, such as the β_2-adrenergic receptor and rhodopsin, are also regulated by phosphorylation/dephosphorylation events (Applebury and Hargrave, 1986; Benovic et al., 1988a). For the β_2-adrenergic receptor several distinct protein kinases have been shown to phosphorylate and regulate the functional activity of the receptors. These include the cAMP-dependent protein kinase (Benovic et al., 1985), protein kinase C (Bouvier et al., 1987), and a recently identified enzyme termed the β-adrenergic receptor kinase or βARK (Benovic et al., 1987a). βARK may, in fact, be a member of a wider family of receptor kinases that share several unique properties, one of which is the agonist-dependent phosphorylation of G protein–coupled receptors (Benovic et al., 1989a). This essay largely deals with the discovery, and structural and functional characterization of the β-adrenergic receptor kinase and its potentially important role in regulating β-adrenergic receptor function. Analogies between this system and the light-activated phosphodiesterase system in the retina are also discussed.

Classification of Desensitization

Adenylyl cyclase–coupled β-adrenergic receptors have served as excellent models for elucidating the molecular basis of receptor desensitization (Sibley and Lefkowitz 1985; Benovic et al., 1988a). Cellular systems containing β-adrenergic receptor-coupled adenylyl cyclase characteristically show a decreased response to agonist stimulation after even brief periods of prior exposure to the agonist. Numerous mechanisms have been shown to contribute to this overall decrease in responsiveness. These mechanisms play out over seconds, minutes, hours, and days. They involve changes in the functional properties, number, and subcellular distribution of the receptors as well as of the other components of the system such as the G proteins and adenylyl cyclase. It was from an attempt to understand the mechanisms operating to rapidly uncouple the receptors from the adenylyl cyclase system that the discovery of βARK emerged. At a phenomenological level rapid desensitization phenomena are often classified as being either heterologous or homologous. Heterologous desensitization refers to a situation where prior exposure of a cell to an activator, such as epinephrine, leads to subsequent refractoriness to a whole variety of activators including those working through receptors other than the β-adrenergic receptor (Sibley and Lefkowitz, 1985). Some cases of heterologous desensitization appear to be mediated by changes in components distal to the receptors. In these systems decreased responsiveness is observed even with activators that bypass the receptors, such as guanine nucleotides, sodium fluoride, forskolin, etc. In contrast, homologous desensitization is a very agonist-specific phenomenon. Thus, after stimulation by a β-agonist, the system will show decreased responsiveness only to β-agonists and not to stimulators that act through their own distinct receptors, such as prostaglandin E_1 or glucagon.

Considerable data over the past several years have demonstrated that at least one major mechanism operating to cause heterologous desensitization of the β-adrenergic receptor is feedback regulation via its phosphorylation by the cAMP-dependent protein kinase. Since any activator that raises cAMP levels will cause this to occur, such desensitization is, of necessity, heterologous.

In contrast, it has been known for some years that the process of homologous desensitization is not a cAMP-mediated one. For example, Green and Clark (1981) originally showed, and others have subsequently confirmed, that homologous desensitization can be documented to occur in variant lines derived from S49 lymphoma cell which lack either the cAMP-dependent protein kinase (kin^-) or the G_s protein α subunit (cyc^-). Moreover Strasser et al. (1986b) demonstrated that homologous desensitization occurring in these variant cells was accompanied by rapid agonist-induced phosphorylation of the β-adrenergic receptors. This finding suggested that a kinase existed in these cells which, in an agonist-dependent and cAMP-independent manner, phosphorylated the β-adrenergic receptors.

Identification and Purification of βARK

Our initial attempts to identify a protein kinase involved in homologous desensitization utilized the kin^- mutant of S49 lymphoma cells. This cell line was chosen for initial study because it exhibits typical homologous (agonist-specific) desensitization and β-adrenergic receptor phosphorylation while lacking the cAMP-dependent protein kinase (Strasser et al., 1986b). When reconstituted hamster lung $β_2$-adrenergic receptor was incubated with kin^- cell crude soluble fraction under phosphorylating conditions low levels of receptor phosphorylation were observed (Fig. 1, lane 1). However, when the receptor was occupied with a β-agonist such as isoproterenol, phosphorylation was enhanced 10-fold (lane 2). Moreover, the agonist effect could be completely blocked by coincubation with a β-antagonist such as alprenolol. These initial studies demonstrated the existence of a kinase activity that specifically phosphorylated the agonist-occupied form of the receptor (Benovic et al., 1986b).

This kinase activity, which was termed the β-adrenergic receptor kinase or βARK, was partially purified from the kin^- cell supernatant and was shown to be distinct from the cAMP-, cGMP-, Ca^{2+}/calmodulin- and Ca^{2+}/phospholipid-dependent protein kinases. Moreover, βARK activity is present in a large number of tissues suggesting that agonist-dependent phosphorylation may serve as a ubiquitous mechanism for receptor regulation.

βARK has been purified from bovine cerebral cortex to >90% homogeneity by sequential chromatography on Ultrogel AcA34, diethylaminoethyl (DEAE) Sephacel, carboxymethyl (CM) Fractogel, and hydroxylapatite (Benovic et al., 1987b). This results in an ~20,000-fold purification with an overall recovery of 12%. The purified kinase consists of a single subunit of 80,000 D which is able to phosphorylate the agonist-occupied β_2-adrenergic receptor to a stoichiometry of 8 mol phosphate/mol receptor. The kinetics of the phosphorylation reaction ($K_m = 0.25$ μM for β-adrenergic receptor, $K_m = 35$μM for ATP, $V_{max} = $ ~80 nmol phosphate/min per mg protein) suggest a relatively high affinity interaction between the kinase and receptor. Once purified, βARK appears to be very labile having a high life of only several days at 4°C. However, the addition of 0.02% Triton X-100 to the buffer dramatically stabilizes the enzyme resulting in preparations that have minimal loss of activity after several months.

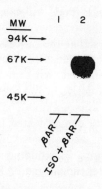

Figure 1. Phosphorylation of purified hamster lung β-adrenergic receptor by the supernatant fraction of lysed kin⁻ cells. Reconstituted receptor (6 pmol) was incubated with a supernatant fraction from high-speed centrifugation of lysed kin⁻ cells (60 μg of protein) for 30 min at 30°C without (lane 1) or with the addition of 10 μM (−)isoproterenol (lane 2). The phosphorylated receptor was then purified before electrophoresis on a 10% polyacrylamide gel. The molecular weight standards are shown times 10^{-3}.

Substrate Specificity of βARK

Several studies have attempted to address the substrate specificity of βARK. These studies have focused on (a) assessing the agonist-dependent nature of the phosphorylation reaction, (b) determining the sites of phosphorylation on the receptor, and (c) determining the receptor specificity for phosphorylation by βARK.

As discussed previously, βARK has a strong preference for phosphorylating the agonist-occupied or activated form of the receptor. This is, in a sense, analogous to the enhanced ability of the G protein G_s to interact with the hormone-occupied vs. unoccupied receptor (Cerione et al., 1984). The agonist-dependent phosphorylation of the receptor by βARK has been characterized by studying the ability of full and partial agonists to promote β_2AR phosphorylation (Benovic et al., 1988b). Partial agonists appear to promote reduced receptor phosphorylation when compared with full agonists. Morever, there is an excellent correlation between the ability of partial agonists to stimulate adenylyl cyclase activity and promote receptor phosphorylation by βARK suggesting that both G_s and βARK interaction with the receptor are enhanced by a common agonist-induced conformational change.

In an effort to determine if the differing abilities of various partial agonists to promote receptor phosphorylation reflects primarily an effect on the rate or on the maximum extent of phosphorylation, time-course experiments were performed. As shown in Fig. 2 isoproterenol-promoted βAR phosphorylation is near maximal at 30 min with a stoichiometry of 9–10 mol phosphate/mol receptor. Phosphorylation of the β-adrenergic receptor in the presence of no ligand or when occupied by the partial agonists Lilly 46220 or dichloroisoproterenol (DCI) is also near maximal at 30 min, however, the stoichiometries are significantly less than that achieved with isoproterenol. These studies thus demonstrate that even when the receptor is fully occupied, partial agonists are only able to promote submaximal receptor phosphorylation by βARK. Moreover, the kinetics of the phosphorylation reactions suggest that the major difference between phosphorylation of agonist- vs. partial agonist-occupied vs. unoccupied receptor is in the rate or extent (V_{max}) at which the receptor is phosphorylated.

The sites of βARK phosphorylation on the mammalian $β_2$AR remain somewhat obscure although several lines of evidence suggest thay are localized to the carboxyl-

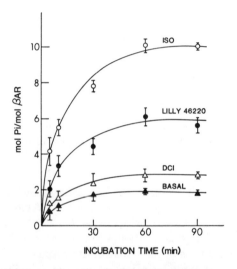

Figure 2. Time course of partial agonist-stimulated phosphorylation of the β-adrenergic receptor by βARK. Purified hamster lung βAR, reinserted into phosphatidylcholine vesicles, was incubated with CM Fractogel purified βARK (~2 μg) in the presence of 20 mM Tris-HCl, pH 7.5, 2 mM EDTA, 4 mM MgCl$_2$, 4 mM sodium phosphate, 4 mM NaF, and 75 μM [$γ^{32}$P]ATP (~2 cpm/fmol). Samples were incubated for the indicated period of time at 30°C with either 20 μM (−)isoproterenol (*ISO*), 300 μM Lilly 46220, 20 μM dichloroisoproterenol (*DCI*) or no ligand (*basal*). Reactions were stopped by the addition of SDS sample buffer followed by electrophoresis on a 10% polyacrylamide gel. Stoichiometries were determined by cutting and counting the dried gel as previously described (Benovic et al., 1986b). The data shown represent the mean ± SE from three to seven individual experiments. In this series of experiments very high levels of βARK were used (~2 μg) so as to evoke maximum phosphorylation even in the presence of partial agonists.

terminal tail of the receptor. Dohlman et al. (1987b) have demonstrated that most, if not all, of the sites on the βARK phosphorylated β-receptor were lost after carboxypeptidase treatment, which removes the carboxyl tail of the receptor. In addition, mutagenesis studies also target the carboxyl tail as a major phosphorylation domain. Using cells expressing a mutant $β_2$-adrenergic receptor lacking ~50 amino acids at the carboxyl terminus, Bouvier et al. (1988) have demonstrated that the mutant receptor does not undergo the agonist-promoted phosphorylation observed for the wild-type receptor. Similar studies, which use a mutant $β_2$-receptor that has the carboxyl-terminal serines and threonines changed to alanines or glycines, also show significantly reduced agonist-dependent phosphorylation (Hausdorff et al., 1989). An additional line of study has used synthetic peptides from various domains of the $β_2$-receptor as potential substrates for βARK (Benovic et al., 1990). Of 11 peptides tested only two were able to serve as substrates for βARK. These two peptides encompassed some 66 amino acids from the carboxyl terminus of the receptor and serve as further evidence that this is the domain phosphorylated by βARK.

While the β-adrenergic receptor-coupled adenylyl cyclase system has proven to be a very useful model for studying agonist-specific desensitization, many other receptors coupled to this system are also susceptible to the same type of desensitization. Thus, it is not unreasonable to speculate that the mechanism by which these other receptors undergo homologous desensitization might also involve an agonist-induced receptor phosphorylation. The notion that each receptor has its own specific protein kinase, however, seems unlikely. A more plausible hypothesis is that a general receptor kinase exists that is able to phosphorylate multiple agonist-occupied receptors. Early in these studies it was found that stimulation of DDT_1 MF-2 hamster vas deferens smooth muscle cells or S_{49} lymphoma cells with a β-agonist leads to the sequestration of up to ~80% of the βARK activity from the cytosol to the plasma membrane (Strasser et al., 1986a). The sequestration was rapid and reversible and coincided temporally with the time courses of receptor phosphorylation and desensitization. This suggested that the agonist-induced sequestion of βARK may represent the first step in homologous desensitization. Moreover, this property provided a test for the hypothesis that βARK can phosphorylate different agonist-occupied receptors. In S_{49} cells, prostaglandin E_1, which induces homologous desensitization to its own actions, also promotes βARK sequestration to the plasma membrane. This result suggests, albeit indirectly, that βARK may be a general adenylyl cyclase-coupled receptor kinase capable of phosphorylating and regulating the function of many receptors.

Similar results were obtained when S_{49} lymphoma cells were treated with somatostatin, a 14 amino acid peptide that inhibits adenylyl cyclase activity through a specific receptor. In these cells, somatostatin induced the sequestration of βARK from the cytosol to the plasma membrane to an extent comparable to β-agonists or prostaglandin E_1 (Mayor et al., 1987). The sequestration process was rapid and reversible and was associated with desensitization of the somatostatin response, which suggests that βARK also phosphorylates and regulates receptors coupled to inhibition of adenylyl cyclase.

To determine more directly if such receptors serve as substrates for βARK, the effects of this kinase on the $α_2$-adrenergic receptor were studied. In this study, purified human platelet $α_2$-adrenergic receptor, reconstituted into phospholipid vesicles, was found to be a very good substrate for βARK (Benovic et al., 1987c). The phosphorylation was dependent on agonist occupancy of the receptor and was completely blocked by coincubation with $α_2$-antagonists. The time course of phosphorylation of the $α_2$-adrenergic receptor by βARK was virtually identical to that observed with the β-adrenergic receptor. In contrast, the $α_1$-adrenergic receptor, which is coupled to stimulation of phosphatidylinositol hydrolysis, did not serve as a substrate for βARK. Recent studies have also demonstrated that βARK is able to phosphorylate the purified chick heart muscarinic acetylcholine receptor to a high stoichiometry in an agonist-dependent fashion, suggesting that βARK or a related kinase may serve to regulate muscarinic acetylcholine receptor function (Kwatra et al., 1989). Taken together these results strongly suggest that βARK is a general adenylyl cyclase-coupled receptor kinase.

Structure of βARK

The structure of βARK has recently been elucidated by isolation of a cDNA encoding this enzyme (Benovic et al., 1989a). This was accomplished by initially subjecting ~50 μg of purified bovine βARK to cyanogen bromide cleavage. The resultant peptides were resolved by reverse phase high pressure liquid chromatography and subjected to gas-phase sequencing. A size-selected bovine brain cDNA library was then screened using synthetic oligonucleotide probes derived from the peptide amino acid sequences. One of the clones isolated using this approach was 3,949 bp in length and contained an open reading frame of 2,067 bp starting with a potential initiation codon. The open

```
1    MADLEAVLADVSYLMAMEKSKATPAARASKKILLPEPSIRSVMQKYLEDRGEVTFEKIFSQKLGY

66   LLFRDFCLKHLEEAKPLVEFYEEIKKYEKLETEEERLVCSREIFDTYIMKELLACSHPFSKSAIEHV

133  QGHLVKKQVPPDLFQPYIEEICQNLRGDVFQKFIESDKFTRFCQWKNVELNIHLTMNDFSVHRIIG

199  RGGFGEVYGCRKADTGKMYAMKCLDKKRIKMKQGETLALNERIMLSLVSTGDCPFIVCMSYAFH

263  TPDKLSFILDLMNGGDLHYHLSQHGVFSEADMRFYAAEIILGLEHMHNRFVVYRDLKPANILLDE

327  HGHVRISDLGLACDFSKKKPHASVGTHGYMAPEVLQKGVAYDSSADWFSLGCMLFKLLRGHSPF

392  RQHKTKDKHEIDRMTLTMAVELPDSFSPELRSLLEGLLQRDVNRRLGCLGRGAQEVKESPFFRS

456  LDWQMVFLQKYPPPLIPPRGEVNAADAFDIGSFDEEDTKGIKLLDSDQELYRNFPLTISERWQQE

521  VAETVFDTINAETDRLEARKKTKNKQLGHEEDYALGKDCIMHGYMSKMGNPFLTQWQRRYFYLF

585  PNRLEWRGEGEAPQSLLTMEEIQSVEETQIKERKCLLLKIRGGKQFVLQCDSDPELVQWKKELRD

650  AYREAQQLVQRVPKMKNKPRSPVVELSKVPLIQRGSANGL
```

Figure 3. Deduced amino acid sequence of bovine brain βARK. The amino acid sequence shown begins at the first ATG in the nucleotide sequence and is numbered on the left-hand side. The residues encompassing the catalytic domain are underlined.

reading frame codes for a protein of 689 amino acids (79,656 D), similar to the size of purified bovine brain βARK. The amino acid sequence of this clone is shown in Fig. 3. Several lines of evidence suggest that the initial ATG in the nucleotide sequence serves as the initiation site for translation. These include a Kozak consensus sequence, a GC-rich region and an open reading frame that codes for a protein of the predicted molecular size. In addition, in vitro translation studies demonstrate that this ATG can, in fact, be used for the initiation of translation.

The overall topology of βARK suggests an amino-terminal domain of ~197 amino acids, a central catalytic domain of ~239 amino acids, and a carboxyl-terminal domain of ~253 amino acids. The function of the amino- and carboxyl-terminal domains is at present unknown, however, one might speculate that at least one, if not both domains might be involved in mediating the specificity of βARK for phosphorylating only agonist-occupied receptors.

The predicted amino acid sequence of βARK reveals that the catalytic domain has substantial homology with those of other protein kinases. Thus, βARK has a glyXglyXXgly$(X)_{16}$lys stretch which is found in all ATP-binding proteins. The first gly in this stretch is at amino acid 198 and serves and serves to delineate an amino-terminal domain of ~197 amino acids, which bears no significant homology with any other currently sequenced proteins. One potentially interesting difference between the catalytic domain of βARK vs. those of other protein kinases is in the sequence asp phe gly which is found in virtually every protein kinase catalytic domain. The βARK sequence contains a leu in place of the phe in this stretch. The conserved arg at amino acid 436 serves to delineate the end of the catalytic domain from a carboxyl-terminal stretch of 253 amino acids. This carboxyl-terminal region bears no significant homology with any other currently sequenced protein.

The homology of the catalytic domain of βARK with other protein kinases is highest with the cAMP-dependent protein kinase (33.1% identity in a 347 amino acid overlap) and with protein kinase C (33.7% identity in a 303 amino acid overlap). The catalytic domain of βARK has been further compared with the catalytic domains of several other protein kinases by construction of a phylogenetic tree (Benovic et al., 1989a). These results suggest that βARK probably represents the first member of a predicted new subfamily of receptor kinases.

To validate that this cDNA encodes βARK activity a mammalian expression plasmid for βARK was constructed by inserting the cDNA into the expression vector pBC12BI. The corresponding vector, termed pBCβARK, was then used to transiently transfect COS-7 cells. Cells transfected with pBC alone have a low basal activity as assessed by the ability of a lysed cell supernatant fraction to phosphorylate light-bleached rhodopsin. This basal activity corresponds to 1.2 pmol Pi/min per mg protein, a value similar to that found in bovine brain (Benovic et al., 1987b). In contrast, COS-7 cells transfected with pBCβARK show a 43-fold increase in βARK activity, which corresponds to a specific activity of 51 pmol/min per mg. The specificity of the expressed kinase was determined with hamster $β_2$-adrenergic receptor and bovine rhodopsin (Benovic et al., 1986a) as potential substrates. The purified $β_2$-AR serves as a good substrate for the kinase preparation in an agonist-dependent fashion. The kinase preparation also phosphorylated rhodopsin in a light-dependent fashion although the $β_2$-AR is clearly the preferred substrate. However, it remains to be determined whether this kinase preferentially phosphorylates the $β_2$-AR as compared with other G protein–coupled receptors.

Northern blot analysis using the βARK cDNA as a probe reveals a mRNA of 4 kb with the highest levels in brain and spleen. Heart and lung have ~40% the signal of brain while ovary and kidney have only ~20%. There was no distinguishable mRNA band observed in liver and muscle. These results suggest that βARK mRNA may be localized in tissues that have a high degree of sympathetic innervation. In particular, brain, spleen, and heart, which have the highest amounts of βARK mRNA, are highly innervated with correspondingly high concentrations of norepinephrine. In contrast, liver and muscle, in which βARK mRNA is not detectable, show much less adrenergic innervation.

Southern blot analysis of bovine genomic DNA digested with Hind III or Sac I reveals multiple hybridizing bands when probed with a 0.7-kb fragment of the βARK cDNA. Our studies demonstrate that these multiple species are not due to the presence of restriction sites within potential introns. Thus, these additional bands probably represent other genes which the βARK probe is recognizing and suggest that βARK is likely to be a member of a multigene family.

Further demonstration that βARK is a member of a multigene family comes from low stringency hybridization studies. The 0.7-kb fragment from the βARK cDNA was labeled with ^{32}P by nick translation and used to rescreen the bovine brain cDNA library. The labeled fragment hybridized with two classes of cDNA clones. One class hybridized strongly and was not removed by high stringency washing (60°C, 0.1 × SSC). When sequenced, these clones were found to be βARK. A second class of clones hybridized weakly and were removed by moderate stringency washing (60°C, 0.2–0.5 × SSC). This second class of clones, when isolated and sequenced, were found to be distinct from βARK. The cDNA, termed pβARKII, has an opening reading frame which encodes a protein of 688 amino acids (79,610 D). A comparison of the βARK II sequence with that of βARK reveals an overall homology of 85.1% at the amino acid level. Moreover, the catalytic domain of the two kinases (amino acids 198–436) are 95.4% identical while the amino- and carboxyl-terminal domains have somewhat lower identity, 81.2% and 77.7%, respectively. These results suggest that βARKII might also phosphorylate receptors in an agonist-dependent fashion, however, its substrate specificity might differ from βARK. Expression studies should help to elucidate the substrate specificity of βARKII and its potential role in agonist-specific desensitization.

Physiological Role of βARK

Several lines of experimental evidence have implicated the involvement of receptor phosphorylation in rapid desensitization of the β-adrenergic and related G protein–coupled receptors. The enzymes involved in these reactions are the cAMP-dependent protein kinase, which mediates heterologous or agonist nonspecific desensitization, and βARK, which is involved in homologous or agonist-specific desensitization. The homologous nature of βARK effects on several different receptors is presumably explained by its unique specificity for selectively phosphorylating the agonist-occupied forms of these receptors. There are several lines of evidence that support a role for βARK in mediating agonist-specific desensitization. When pure β-adrenergic receptor is phosphorylated by pure βARK in vitro, the resulting phosphorylated receptor has only a slightly reduced ability to interact with G_s (16% inactivation), as assessed by agonist-stimulated GTPase activity in a reconstituted system (Benovic et al., 1987a). However, when pure β-adrenergic receptor is phosphorylated with a crude βARK preparation (~1% pure), the receptor interaction with G_s is significantly reduced (~80% inactivation). This suggests that the cruder βARK fraction contains a factor that enhances the uncoupling of phosphorylated β-adrenergic receptor and G_s. Moreover, the addition of pure retinal arrestin (Wilden et al., 1986) to β-adrenergic receptor phosphorylated by pure βARK enhanced the uncoupling of phosphorylated β-adrenergic receptor and G_s (41% inactivation) (Benovic et al., 1987a) (Fig. 4).

Arrestin (also called 48-kD protein or S antigen) is involved in enhancing the inactivating effect of rhodopsin phosphorylation by rhodopsin kinase (see section on *Regulation of the Visual System* below). These results suggest that a protein analogous to retinal arrestin may be involved in mediating the functional consequences of receptor phosphorylation by βARK.

Bouvier et al. (1988), using cells expressing a mutant $β_2$-adrenergic receptor that lacks the serine- and threonine-rich carboxyl terminus, showed that such a mutated receptor does not undergo the agonist-promoted phosphorylation observed in cells expressing the wild-type receptor. This observation strongly implicates βARK as the enzyme responsible for the agonist-induced phosphorylation of the β-receptor in whole cells. This study also implicates βARK in some aspect of agonist-specific (homologous) desensitization. Indeed, cells expressing the carboxyl-terminal truncated receptor (which did not undergo agonist-induced phosphorylation) display a delayed onset of

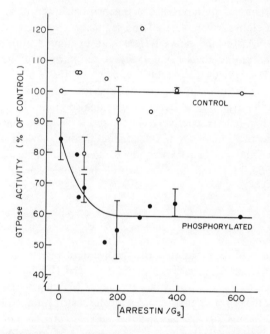

Figure 4. Effect of retinal arrestin on inhibiting $βAR$-G_s coupling. Phospholipid vesicles containing purified βAR were incubated for 60 min with purified βARK (>90%). The stoichiometry of phosphorylation was ≈8 mol of phosphate per mol of receptor. Control (p[NH]ppA) and phosphorylated (ATP) βAR preparations (≈0.4 pmol) were then incubated with purified retinal arrestin (0–1,500 pmol) and pure G_s (≈2.4 pmol) for 20 min at 4°C. Samples were then assayed for isoproterenol-stimulated GTPase activity. The results are data from five separate experiments. The error bars represent means ± SEM from two to five independent experiments; individual points are data from single experiments.

agonist-induced desensitization. Such mutated receptors, however, showed a normal desensitization pattern after longer agonist exposure (Strader et al., 1987; Bouvier et al., 1988). A normal (Strader et al., 1987), as well as an increased (Bouvier et al., 1988) sequestration of the receptor has been reported in cells expressing the truncated receptor. Moreover, in another form of the receptor in which carboxyl-terminal serine and threonine residues have been mutated, desensitization is slowed even further (Bouvier et al., 1988; Hausdorff et al., 1989). Agonist-promoted phosphorylation of the carboxyl terminus of the receptor, presumably by βARK, therefore appears to be a crucial event in the early stages of homologous desensitization.

A third line of evidence for the involvement of βARK in desensitization comes from studies using protein kinase inhibitors. Polyanions such as heparin and dextran sulfate are potent inhibitors of βARK with K_i's of 10–50 μM (Benovic et al., 1989b). The inhibition of heparin appears to be competitive with the substrate. The ability of heparin to potently inhibit receptor phosphorylation by βARK has been used in a permeabilized cell system to address the role of βARK in desensitization (Lohse et al., 1989, 1990). These studies demonstrate that inhibitors of βARK are able to markedly inhibit agonist-induced receptor phosphorylation and homologous desensitization. Moreover, under conditions where only homologous desensitization is being measured, inhibitors of the cAMP-dependent protein kinase and of protein kinase C had no effect on the desensitization.

The ability of the cAMP-dependent kinase and βARK to mediate rapid desensitization occurs over quite different ranges of agonist concentrations. Since stimulation of cAMP-dependent kinase activity is approaching maximal levels at very low concentrations of catecholamine (10^{-8} M), desensitization mediated by this mechanism occurs at these low concentrations. In contrast, βARK-mediated phosphorylation occurs only with significant levels of receptor occupancy (typically 10^{-8} to 10^{-6} M catecholamine). These findings might suggest that the cAMP-dependent protein kinase is primarily involved in mediating desensitization at peripheral receptors responding to the very low circulating levels of epinephrine. In contrast, βARK would seem better adapted to respond to the much higher neurotransmitter levels found at sympathetic synapses.

There are many unanswered questions about βARK and other putative related gene products. How many such kinases are there? What is their specificity and distribution? Rhodopsin kinase is an obvious candidate as are ligand-dependent receptor kinases in *Dictyostelium discoideum* and yeast. Understanding the structure and function of this family of receptor kinases should provide fundamental insights into one of the most pervasive and basic types of biological phenomena, regulation of cellular sensitivity to environmental stimuli.

Regulation of the Visual System

Analogous to the β-adrenergic receptor, rod cells of the retina contain a G protein–linked signal transduction pathway. In this system, rhodopsin, a 40,000 D integral membrane glycoprotein with a seven-transmembrane-spanning topography, functions as a receptor for a photon of light (for reviews: Applebury and Hargrave, 1986; Nathans, 1987). Rhodopsin contains a molecule of 11-*cis* retinal bound via a protonated Schiff's base to lysine 296 in the seventh transmembrane-spanning region of the photoreceptor (Findlay et al., 1981). Absorption of a photon of light causes the isomerization of 11-*cis* retinal to the all-*trans* form of the chromophore, which results in a conformational change in the protein. When photoactivated, rhodopsin interacts with several cytosolic proteins including transducin, a guanine nucleotide–binding protein present in the retina. This interaction promotes the exchange of GDP for GTP on the α subunit (T_α) of tranducin (Fung and Stryer, 1980) resulting in activation of a cGMP phosphodiesterase by removal of an inhibitory subunit (Hurley and Stryer, 1982).

Since the half-life of the photointermediate (metarhodopsin II) which activates transducin is long-lived (Kuhn and Wilden, 1987) and the decay to metarhodopsin III is delayed by transducin until T_α-GTP dissociates from the photoreceptor (Pfister et al., 1983) an efficient means of terminating the signal exists. This mechanism involves two retinal cytosolic proteins: rhodopsin kinase and arrestin. Phosphorylation of light-activated rhodopsin by rhodopsin kinase is a key step in the signal termination reaction. In vitro, phosphorylated rhodopsin is less effective in stimulating phosphodiesterase activity (Sitaramayya and Liebman, 1983a, b; Aton and Litman, 1984; Wilden et al., 1986). After phosphorylation by rhodopsin kinase, the photoreceptor now binds arrestin (also referred to as 48 K-protein or S-antigen) which sterically hinders transducin activation (Kuhn et al., 1985). Arrestin markedly potentiates the effects of phosphorylated Rho* while the addition of arrestin to nonphosphorylated Rho* has no effect on phosphodiesterase activity (Wilden et al., 1986). By the action of rhodopsin kinase and arrestin, the photoactivation pathway is turned off long before Rho* decays to a form that will not interact with transducin.

Rhodopsin represents the first membrane receptor for which signal-dependent phosphorylation was observed (Bownds et al., 1982; Kuhn and Dreyer, 1972). Kuhn et al. (1973) later suggested that the light-dependent phosphorylation event involved alterations of the substrate, rhodopsin, and not photoactivation of the kinase directly. The phosphorylation of Rho* by rhodopsin kinase results in the formation of multiple species of phosphorylated rhodopsin with stoichiometries of 1–9 mol phosphate/mol rhodopsin. The number of sites phosphorylated could be increased by conditions that favor regeneration and rebleaching of the substrate, or by increasing the ratio of kinase/Rho* with the addition of soluble rhodopsin kinase (Wilden and Kuhn, 1982). Numerous investigators have localized the sites of phosphorylation to the carboxy-terminal portion of the molecule (Sale et al., 1978; Pellicone et al., 1981; Hargrave et al., 1982). Thompson and Findley (1984) used peptide sequencing in concert with subdigestion procedures to directly identify five sites of phosphorylation including serines 334, 338, and 343 as well as threonines at 335 and 336 in the ovine rhodopsin molecule. Additionally, it has been suggested that serine 240, located in a cytoplasmic loop connecting transmembrane-spanning helices V and VI, serves as a site of phosphorylation of rhodopsin kinase (Applebury and Hargrave, 1986).

Rhodopsin kinase has been purified by standard chromatographic procedures (Shichi and Somers, 1978) in addition to the use of light-exposed rod outer segment membranes which serve as an "affinity matrix" for the kinase (Palczewski et al., 1988a; Okada and Ikai, 1988). The later purification schemes exploit the property of rhodopsin kinase that it binds tightly to photoactivated rod outer segment membranes, which requires higher ionic strength conditions to extract the kinase than if dark-adapted rod outer segments are used. A low ionic strength wash is first used to remove contaminating proteins and a higher ionic strength wash (60 mM KCl) resulted in an 80-fold purification with a yield of 60–80%. The kinase was then purified to homogeneity by ion exchange chromatography on DEAE-cellulose followed by a hydroxyapatite step. The overall recovery approached 20% and the purified kinase had an apparent molecular weight of 67,000 by SDS-PAGE (Palczewski et al., 1988a), which is in good agreement with the molecular weight observed by molecular sieve chromatography on Sephadex G-100. A 50,000 molecular weight species possesses rhodopsin kinase activity (Shichi and Somers, 1978), however, this moiety appears to be a proteolytic fragment derived from the 67,000 D protein (Palczewski et al., 1988a). Once purified the kinase is extremely labile with reports that 15% glycerol (Shichi et al., 1983) or 20% adonitol (Palczewski et al., 1988a) help to preserve activity. The enzyme activity is markedly inhibited by detergent (Shichi et al., 1983) with the exception of dodecyl maltoside (6 mM) which has negligible effects on kinase activity at 15 min (Palczewski et al., 1988b).

The rate of phosphorylation of Rho* by rhodopsin kinase follows pseudo first-order kinetics ($k = 0.06$ min^{-1}) when the ATP concentration is 30 times that of rhodopsin (Shichi and Somers, 1978). The kinase has a marked preference for ATP to serve as the phosphate donor with a $K_m = 4\ \mu M$ and a $V_{max} = 700$ nmol/min per mg as compared to GTP with a $K_m = 1$ mM and a $V_{max} = 10$ nmol/min per mg (Palczewski et al., 1988a). Rhodopsin kinase is competitively inhibited by adenosine, AMP, and ADP but most other adenosine derivatives as well as cyclic nucleotides have no significant effect on the kinase activity (Shichi and Somers, 1978; Lee et al, 1982; Palczewski et al., 1988a, b).

As mentioned above, rhodopsin kinase is highly specific for photoactivated rhodopsin to serve as substrate. In vitro experiments performed in the dark exhibit low levels of phosphate incorporation, which appear to be due to residual amounts of bleached rhodopsin in the preparation. This level of phosphorylation can be abolished by using chromatographically pure rhodopsin free of any opsin (Palczewski et al., 1988b). Once the photointermediate metarhodopsin II is formed, the photoreceptor's ability to serve as a substrate for the kinase decays with a $t_{1/2}$ of ~30 min (Miller et al., 1977), which suggests that intermediates beyond metarhodopsin II can be phosphorylated by the kinase (Kuhn and Wilden, 1987). After the formation of metarhodopsin II, phosphorylation is not effected by changes in the conformation of the substrate since regeneration of rhodopsin by the addition of 11-cis retinal in the dark has no effect on the phosphorylation reaction (Miller and Paulsen, 1975). The agonist-occupied form of the β_2-adrenergic receptor also serves as a substrate, albeit a poor one when compared to its interaction with β-adrenergic receptor kinase (Benovic et al., 1986a), which demonstrates that rhodopsin kinase can phosphorylate other G protein–coupled receptors in an agonist-dependent manner. Rhodopsin kinase undergoes autophosphorylation independent of light or the presence of cyclic nucleotides (Lee et al., 1982). However, common kinase substrates such as histones, casein, phosvitin, and protamine do not serve as substrates for the enzyme (Shichi et al., 1983). Additionally, rhodopsin kinase is not activated by cyclic nucleotides, phosphoinositols, nor cations and does not require phospholipid for activity (Palczewski et al., 1988a, b).

The question arises as to what is the mechanism by which rhodopsin kinase phosphorylates Rho* in a signal-dependent fashion. Synthetic peptides whose sequences are based on known phosphorylation sites from the intact photoreceptor serve as substrates for the kinase (Palczewski et al., 1988b, 1989) with K_m's ~1,000-fold greater than that of the intact protein (rhodopsin K_m ~3 μM). This result suggests that the interaction of the kinase and rhodopsin is not defined purely by the primary or secondary structure of the phosphorylation site. It is hypothesized that the interaction of the kinase and the photoactivated substrate requires multiple sites of interaction on the cytoplasmic surface of Rho* and this is reflected by the difference in K_m's for the phosphorylation reactions. In support of this model, peptides whose sequences represent the third cytoplasmic loop (residues 231–252) or a portion of the carboxy-terminal region (337–348) proved to be inhibitors of Rho* phosphorylation (Palczewski et al., 1988b) while the third-loop peptide has no effect on the phosphorylation of the carboxy-terminal peptide (324–348) substrate (Palczewski et al., 1989). Recent work by Konig et al. (1989) provides evidence that three intracellular regions of rhodopsin interact with transducin based on the ability of synthetic peptides to compete with metarhodopsin II for binding with transducin. The formation of the intermediate metarhodopsin II must result in a conformational change which permits both transducin and rhodopsin kinase to interact with portions of the cytoplasmic surface previously inaccessible to either protein.

Data obtained with synthetic peptides indicate that acidic groups may play a role in substrate recognition by rhodopsin kinase (Palczewski et al., 1989). Peptides with sequences based on portions of the red and green visual pigments, the hamster β_2-adrenergic receptor and the M_1-muscarinic receptor all serve as substrates for rhodopsin kinase as did an acidic peptide from casein. Basic peptides were not substrates for rhodopsin kinase. Polycations such as spermine served to activate the kinase as judged by the phosphorylation of Rho* or a synthetic peptide substrate. Polyanions such as polyaspartic acid and dextran sulfate inhibit rhodopsin kinase. The inhibition of poly-(1-aspartic acid) was found to be competitive with rhodopsin ($K_i = 300$ μM), which suggests the importance of negative charge in rhodopsin kinase interactions.

In summary, the rhodopsin system provides an excellent model for studying the molecular mechanisms of G protein–mediated signal transduction. Rhodopsin kinase, an enzyme with properties very similar to βARK, plays a central role in terminating the signal that results from photoactivation of rhodopsin. The extreme substrate specificity and the ability to phosphorylate receptors in an agonist-dependent manner, is necessitated by the biologic function of this family of kinases.

References

Applebury, M. L., and P. A. Hargrave. 1986. Molecular biology of the visual pigments. *Vision Research.* 26:1881–1895.

Aton, G. B., and B. J. Litman. 1984. Activation of rod outer segment phosphodiesterase by enzymatically altered rhodopsin, a regulatory role for the carboxy terminus of rhodopsin. *Experimental Eye Research.* 38:547–559.

Benovic, J. L., M. Bouvier, M. G. Caron, and R. J. Lefkowitz. 1988a. Regulation of adenylyl cyclase-coupled β-adrenergic receptors. *Annual Review of Cell Biology.* 4:405–428.

Benovic, J. L., C. Staniszewski, F. Mayor, Jr., M. G. Caron, and R. J. Lefkowitz. 1988b. β-Adrenergic receptor kinase: activity of partial agonists for stimulation of adenylate cyclase correlates with ability to promote receptor phosphorylation. *Journal of Biological Chemistry.* 263:8856–8858.

Benovic, J. L., A. DeBlasi, W. C. Stone, M. G. Caron, and R. J. Lefkowitz. 1989a. β-Adrenergic receptor kinase: Primary structure delineates a multigene family. *Science.* 246:235–240.

Benovic, J. L., W. C. Stone, M. G. Caron, and R. J. Lefkowitz. 1989b. Inhibition of the β-adrenergic receptor kinase by polyanions. *Journal of Biological Chemistry.* 264:6707–6710.

Benovic, J. L., H. Kuhn, I. Weyland, J. Codina, M. G. Caron, and R. J. Lefkowitz. 1987a. Functional desensitization of the isolated β-adrenergic receptor by the β-adrenergic receptor kinase: potential role of an analog of the retinal protein arrestin (48 kDa protein). *Proceedings of the National Academy of Sciences.* 84:8879–8882.

Benovic, J. L., F. Mayor, Jr., C. Staniszewski, R. J. Lefkowitz, and M. G. Caron. 1987b. Purification and characterization of the β-adrenergic receptor kinase. *Journal of Biological Chemistry.* 262:9026–9032.

Benovic, J. L., J. W. Regan, H. Matsui, F. Mayor, Jr., S. Cotecchia, L. M. F. Leeb-Lundberg, M. G. Caron, and R. J. Lefkowitz. 1987c. Agonist-dependent phosphorylation of the α_2-adrenergic receptor by the β-adrenergic receptor kinase. *Journal of Biological Chemistry.* 262:17251–17253.

Benovic, J. L., F. Mayor, Jr., R. L. Somers, M. G. Caron, and R. J. Lefkowitz. 1986*a*. Light-dependent phosphorylation of rhodopsin by β-adrenergic receptor kinase. *Nature.* 321:869–872.

Benovic, J. L., R. H. Strasser, M. G. Caron, and R. J. Lefkowitz. 1986*b*. β-Adrenergic receptor kinase: identification of a novel protein kinase which phosphorylates the agonist-occupied form of the receptor. *Proceedings of the National Academy of Sciences.* 83:2797–2801.

Benovic, J. L., J. Onorato, M. J. Lohse, H. G. Dohlman, C. Staniszewski, M. G. Caron, and R. J. Lefkowitz. 1990. Synthetic peptides of the hamster β_2-adrenergic receptor as substrates and inhibitors of the β-adrenergic receptor kinase. *British Journal of Clinical Pharmacology.* In press.

Benovic, J. L., L. J. Pike, R. A. Cerione, C. Staniszewski, T. Yoshimasa, J. Codina, M. G. Caron, and R. J. Lefkowitz. 1985. Phosphorylation of the mammalian β-adrenergic receptor by cyclic AMP-dependent protein kinase: regulation of the rate of receptor phosphorylation and dephosphorylation by agonist occupancy and effects on coupling of the receptor to the stimulatory guanine nucleotide regulatory protein. *Journal of Biological Chemistry.* 260:7094–7101.

Bouvier, M., W. P. Hausdorff, A. DeBlasi, B. F. O'Dowd, B. K. Kobilka, M. G. Caron, and R. J. Lefkowitz. 1988. Removal of phosphorylation sites from the β_2-adrenergic receptor delays onset of agonist-promoted desensitization. *Nature.* 333:370–373.

Bouvier, M., L. M. F. Leeb-Lundberg, J. L. Benovic, M. G. Caron, and R. J. Lefkowitz. 1987. Regulation of adrenergic receptor function by phosphorylation: II. Effects of agonist occupancy on phosphorylation of α_1- and β-adrenergic receptors by protein kinase C and the cyclic AMP-dependent protein kinase. *Journal of Biological Chemistry.* 262:3106–3113.

Bownds, D., J. Dawes, J. Miller, and M. Stahlmann. 1972. Phosphorylation of frog photoreceptor membranes induced by light. *Nature.* 237:125–127.

Cerione, R. A., J. Codina, J. L. Benovic, R. J. Lefkowitz, L. Birnbaumer, and M. G. Caron. 1984. The mammalian β_2-adrenergic receptor: reconstitution of functional interactions between the pure receptor and the pure stimulatory nucleotide binding protein (N_s) of the adenylate cyclase system. *Biochemistry.* 23:4519–4525.

Dohlman, H. G., M. Bouvier, J. L. Benovic, M. G. Caron, and R. J. Lefkowitz. 1987*a*. A family of receptors coupled to guanine nucleotide regulatory proteins. *Biochemistry.* 26:2657–2664.

Dohlman, H. G., M. Bouvier, J. L. Benovic, M. G. Caron, and R. J. Lefkowitz. 1987*b*. The multiple membrane spanning topography of the β_2-adrenergic receptor: localization of the sites of binding, glycosylation and regulatory phosphorylation by limited proteolysis. *Journal of Biological Chemistry.* 262:14282–14288.

Findlay, J. B. C., M. Brett, and D. J. C. Pappin. 1981. Primary structure of C-terminal functional sites in ovine rhodopsin. *Nature.* 293:314–316.

Fung, B. K.-K., and L. Stryer. 1980. Photolyzed rhodopsin catalyzes the exchange of GTP for bound GDP in retinal rod outer segments. *Proceedings of the National Academy of Sciences.* 77:2500–2504.

Green, D. A., and R. B. Clark. 1981. Adenylate cyclase coupling proteins are not essential for agonist-specific desensitization of lymphoma cells. *Journal of Biological Chemistry.* 256:2105–2108.

Hargrave, P. A., J. H. McDowell, E. C. Siemiatkowski-Juszczak, S.-L. Fong, H. Kuhn, J. K. Wang, D. R. Curtis, J. K. M. Rao, P. Argos, and R. J. Feldman. 1982. The carboxy-terminal one-third of bovine rhodopsin: its structure and function. *Vision Research.* 22:1429–1438.

Hausdorff, W. P., M. Bouvier, B. F. O'Dowd, G. P. Irons, M. G. Caron, and R. J. Lefkowitz. 1989. Phosphorylation sites on two domains of the β_2-adrenergic receptor are involved in distinct pathways of receptor desensitization. *Journal of Biological Chemistry*. 264:12657–12665.

Hurley, J. B., and L. Stryer. 1982. Purification and characterization of the gamma regulatory subunit of the cyclic GMP phosphodiesterase from retinal rod outer segments. *Journal of Biological Chemistry*. 257:11094–11099.

Konig, B., A. Arendt, J. H. McDowell, M. Kahlert, P. A. Hargrave, and K. P. Hofman. 1989. Three cytoplasmic loops of rhodopsin interact with transducin. *Proceedings of the National Academy of Sciences*. 86:6878–6882.

Kuhn, H., J. H. Cook, and W. J. Dreyer. 1973. Phosphorylation of rhodopsin in bovine photoreceptor membranes. A dark reaction after illumination. *Biochemistry*. 12:2495–2502.

Kuhn, H., and W. J. Dreyer. 1972. Light-dependent phosphorylation of rhodopsin by ATP. *FEBS Letters*. 20:1–6.

Kuhn, H., S. W. Hall, and U. Wilden. 1985. Light-induced binding of 48-kDa protein to photoreceptor membranes is highly enhanced by phosphorylation of rhodopsin. *FEBS Letters*. 176:473–478.

Kuhn, H., and U. Wilden. 1987. Deactivation of photoactivated rhodopsin by rhodopsin-kinase and arrestin. *Journal of Receptor Research*. 7:283–293.

Kwatra, M. M., J. L. Benovic, M. G. Caron, R. J. Lefkowitz, and M. M. Hosey. 1989. Phosphorylation of chick heart muscarinic cholinergic receptors by the β-adrenergic receptor kinase. *Biochemistry*. 28:4543–4547.

Lee, B. H., B. M. Brown, and R. N. Lolley. 1982. Autophosphorylation of rhodopsin kinase from retinal rod outer segments. *Biochemistry*. 21:3303–3307.

Lohse, M. J., J. L. Benovic, M. G. Caron, and R. J. Lefkowitz. 1990. Multiple pathways of rapid β_2-adrenergic receptor desensitization: delineation with specific inhibitors. *Journal of Biological Chemistry*. 265:3202–3209.

Lohse, M. J., R. J. Lefkowitz, M. G. Caron, and J. L. Benovic. 1989. Inhibition of β-adrenergic receptor kinase prevents rapid homologous desensitization of β_2-adrenergic receptors. *Proceedings of the National Academy of Sciences*. 86:3011–3015.

Mayor, F., Jr., J. L. Benovic, M. Caron, and R. J. Lefkowitz. 1987. Somatostatin induces translocation of the β-adrenergic receptor kinase and desensitizes somatostatin receptors in S_{49} lymphoma cells. *Journal of Biological Chemistry*. 262:6468–6471.

Miller, J. A., and R. Paulsen. 1975. Phosphorylation and dephosphorylation of frog rod outer segment membranes as part of the visual process. *Journal of Biological Chemistry*. 250:4427–4432.

Miller, J. A., R. Paulsen, and M. D. Bownds. 1977. Control of light-activated phosphorylation in frog photoreceptor membranes. *Biochemistry*. 16:2633–2639.

Nathans, J. 1987. Molecular biology of visual pigments. *Annual Review of Neuroscience*. 10:163–194.

Okada, D., and A. Ikai. 1988. Purification method of bovine rhodopsin kinase using regeneration of rhodopsin. *Analytical Biochemistry*. 169:428–431.

Palczewski, K., A. Arendt, J. H. McDowell, and P. A. Hargrave. 1989. Substrate recognition determinants for rhodopsin kinase: studies with synthetic peptides, polyanions and polycations. *Biochemistry*. 28:8764–8770.

Palczewski, K., J. H. McDowell, and P. A. Hargrave. 1988a. Purification and characterization of rhodopsin kinase. *Journal of Biological Chemistry.* 263:14067–14073.

Palczewski, K., J. H. McDowell, and P. A. Hargrave. 1988b. Rhodopsin kinase: substrate specificity and factors that influence activity. *Biochemistry.* 27:2306–2313.

Pellicone, C., N. Virmaux, G. Nullans, and P. Mandel. 1981. Chemical cleavage of bovine rhodopsin at tryptophanyl bonds: characterization of the proteolytic fragments and the phosphorylated site. *Biochimie.* 63:197–209.

Pfister, C., H. Kuhn, and M. Chambre. 1983. Interaction between photoexcited rhodopsin and peripheral enzymes in frog retinal rods. Influence on the postmetarhodopsin II decay and phosphorylation of rhodopsin. *European Journal of Biochemistry.* 136:489–499.

Sale, G. J., P. Towner, and M. Akhtar. 1978. Topography of the rhodopsin molecule. Identification of the domain phosphorylated. *Biochemistry Journal.* 175:421–430.

Shichi, H., and R. L. Somers. 1978. Light-dependent phosphorylation of rhodopsin. Purification and properties of rhodopsin kinase. *Journal of Biological Chemistry.* 253:7040–7046.

Shichi, H., R. L. Somers, and K. Yamamoto. 1983. Rhodopsin kinase. *Methods in Enzymology.* 99:362–366.

Sibley, D. R., and R. J. Lefkowitz. 1985. Molecular mechanisms of receptor desensitization using the β-adrenergic receptor-coupled adenylate cyclase system as a model. *Nature.* 317:124–129.

Sitaramayya, A., and P. A. Liebman. 1983a. Mechanism of ATP quench of phosphodiesterase activation in rod disc membranes. *Journal of Biological Chemistry.* 258:1205–1209.

Sitaramayya, A., and P. A. Liebman. 1983b. Phosphorylation of rhodopsin and quenching of cyclic GMP phosphodiesterase activation by ATP at weak bleaches. *Journal of Biological Chemistry.* 258:12106–12109

Strader, C. D., I. S. Sigal, A. D. Blate, A. H. Cheung, R. B. Register, E. Rands, B. A. Zemcik, M. R. Candelore, and R. A. F. Dixon. 1987. The carboxyl terminus of the hamster β-adrenergic receptor expressed in mouse L cells is not required for receptor sequestration. *Cell.* 49:855–863.

Strasser, R. H., J. L. Benovic, M. Caron, and R. J. Lefkowitz. 1986a. β-agonist and prostaglandin E_1-induced translocation of the β-adrenergic receptor kinase: evidence that the kinase may act on multiple adenylyl cyclase coupled receptors. *Proceedings of the National Academy of Sciences.* 83:6362–6366.

Strasser, R. H., D. R. Sibley, and R. J. Lefkowitz. 1986b. A novel catecholamine activated adenosine cyclic 3',5'-phosphate independent pathway for β-adrenergic receptor phosphorylation in wild type and mutant S49 lymphoma cells: mechanism of homologous desensitization of adenylyl cyclase. *Biochemistry* 25:1371–1377.

Thompson, P., and J. B. C. Findlay. 1984. Phosphorylation of bovine rhodopsin. Identification of the phosphorylated sites. *Biochemistry Journal.* 220:773–780.

Wilden, U., S. W. Hall, and H. Kuhn. 1986. Phosphodiesterase activation by photoexcited rhodopsin is quenched when rhodopsin is phosphorylated and binds the 48 kDa protein of rod outer segments. *Proceedings of the National Academy of Sciences.* 83:1174–1178.

Wilden, U., and H. Kuhn. 1982. Light-dependent phosphorylation of rhodopsin: number of phosphorylation sites. *Biochemistry.* 21:3014–3022.

Chapter 8

Molecular Genetics of Signal Transduction by Muscarinic Acetylcholine Receptors

Mark R. Brann, Jürgen Wess, and S. V. Penelope Jones

Laboratory of Molecular Biology, National Institute of Neurological Disorders and Stroke, and Receptor Genetics, Inc., Bethesda, Maryland 20892

Introduction

Acetylcholine is perhaps the most ubiquitous and best characterized neurotransmitter. Historically, acetylcholine receptors have been divided into nicotinic and muscarinic receptors because of their selective stimulation by the alkaloids nicotine and muscarine, respectively. Functionally, muscarinic and nicotinic receptors can be differentiated by the mechanism and speed by which their cellular signals are transduced. Nicotinic receptors have within their structure a nonspecific cation conductance. Thus, they are able to rapidly and directly transduce depolarizing signals across the plasma membrane (Stroud and Finer-Moore, 1985). On the other hand, muscarinic receptors mediate a wide diversity of responses that are all slower than those mediated by nicotinic receptors. Muscarinic responses are slower because transduction of their cellular signals requires interaction with additional proteins. For example, all muscarinic responses are likely to involve signal transducing guanine nucleotide–binding proteins (G proteins), and some responses are indirect due to effects on second messenger systems. Second messenger responses known to be activated by muscarinic receptors include inhibition of adenylyl cyclase, stimulation of phosphatidyl inositol (PI) and arachidonic acid metabolism, and release of calcium from cytosolic stores (Harden et al., 1986; Nathanson, 1987).

Muscarinic receptors also activate a wide variety of electrophysiological responses. These responses can be divided into hyperpolarizing and depolarizing activity. Hyperpolarizing responses are found in secretory cells such as in lacrimal or salivary glands (Evans and Marty, 1986; Gallacher and Morris, 1987). These responses are induced by muscarinic activation of calcium-dependent potassium and chloride conductances. In various brain nuclei, muscarinic receptor-induced hyperpolarizations are due to activation of other potassium conductances (Egan and North, 1986; McCormick and Prince, 1986; 1987; McCormick and Pape, 1988), and in heart by the activation of an inwardly rectifying potassium channel (Sakmann et al., 1983; Simmons and Hartzell, 1987). Muscarinic receptors induce depolarizing responses by inhibition of several potassium conductances. For example, muscarinic agonists depolarize cells by inhibiting a non–voltage-dependent resting conductance (Madison et al., 1987), or by inhibiting the m-current, a voltage-gated conductance (Constanti and Sim, 1987). Other potassium conductances inhibited by muscarinic agonists include the conductance underlying the after-hyperpolarization (I_{AHP}), which is activated by calcium entry during an action potential (Müller and Misgeld, 1986; Madison et al., 1987; Galligan et al., 1988). In addition to potassium channels, muscarinic agonists also modulate calcium channels, causing both increases (Clapp et al., 1987; Vivaudou et al., 1988) and decreases in calcium current amplitude (Wanke et al., 1987; Gähwiler and Brown, 1987; Toselli and Lux, 1989).

The wide diversity of physiological effects mediated by muscarinic receptors led to the suggestion that a heterogeneity of muscarinic receptors may exist. This suggestion was supported by the discovery of drugs that are able to discriminate among muscarinic receptors in different tissues. For example, the muscarinic antagonist pirenzepine has higher affinity for receptors located in cerebral cortex than for those located in heart and salivary glands. On the other hand, the muscarinic antagonists AF-DX 116 and hexahydrosiladifenidol have higher affinities for receptors located in heart and salivary glands, respectively (Mutschler et al., 1988). In addition, muscarinic receptors located in these tissues can be functionally distinguished. In heart, muscarinic receptors potently inhibit the enzyme adenylyl cyclase and activate potassium channels via a pertussis toxin (PTX)-sensitive G protein (Pfaffinger et al., 1985; Kirsch et al., 1988), but have a limited ability to stimulate PI metabolism. On the other hand, muscarinic receptors in cerebral cortex and glandular tissues are able to stimulate PI metabolism and open calcium-dependent channels via PTX-insensitive G proteins (Evans and Marty, 1986; Nathanson, 1987).

Studies of clonal cell lines have provided more direct evidence for a heterogeneity of muscarinic receptors based on functional responses. As in heart, muscarinic receptors in NG108-15 neuroblastoma cells are able to inhibit adenylyl cyclase by interaction with a PTX-sensitive G protein. In these cells muscarinic receptors have no effect on PI metabolism, while bradykinin receptors stimulate PI metabolism by a PTX-insensitive G protein. On the other hand, muscarinic receptors present in SK-N-SH neuroblastoma and N132N1 astrocytoma cells are able to stimulate PI metabolism by interaction with PTX-insensitive G proteins. In these cells muscarinic receptors are unable to inhibit adenylyl cyclase while other neurotransmitter receptors do (Harden et al., 1986; Baumgold, 1988).

Genetic Identification of Muscarinic Receptor Subtypes

Muscarinic receptors have been purified from porcine cerebral cortex (Haga and Haga, 1985) and atrium (Peterson et al., 1984). The receptors were subsequently cloned and the cDNA sequences were obtained (Kubo et al., 1986, Peralta et al., 1987b). This work demonstrated that the cortical and atrial receptors are highly related but distinct proteins, and that they show significant sequence homology to both the visual pigment rhodopsin and the β-adrenergic receptor. A common feature all these proteins share is that they mediate signal transduction by interaction with signal-transducing G proteins.

Since there was pharmacological and functional evidence for additional muscarinic receptor subtypes, and since the muscarinic receptors belong to a homologous gene family, these receptor cDNAs were good candidates to be used as templates for cloning of genetically related receptors. Using homology cloning, the human and rat forms of these receptors, as well as three related muscarinic receptors were cloned (Bonner et al., 1987, 1988; Peralta et al., 1987a). The receptors have been designated m1–m5 based on the chronological order in which they have been identified (Bonner et al., 1988).

The locations of the mRNAs that encode each of the five muscarinic receptors have been examined using Northern blot analysis and in situ hybridization histochemistry. As expected, the receptors cloned from cerebral cortex (m1) and atrium (m2) are highly abundant in cerebral cortex and atrium, respectively (Kubo et al., 1986; Peralta et al., 1987b). The m3 receptor is abundant within glandular tissue and smooth muscle (Maeda et al., 1988), and the m1–m5 receptors are differentially expressed in many brain regions (Brann et al., 1988a; Buckley et al., 1988; Weiner and Brann, 1989).

Within cell lines, the receptor mRNAs are also differentially expressed. In NG108-15 cells (Peralta et al., 1987a) only m4 is detected, and in N132N1 (Peralta et al., 1988) and SK-N-SH cells (unpublished observations) only m3 is detected. Overall, the distribution of the mRNAs in different tissues and cell lines correlates well with the functional and pharmacological data which suggests that these different tissues and cells express different muscarinic receptors.

The pharmacological properties of the cloned muscarinic receptors have been examined after expression by both *Xenopus* oocytes and mammalian cells in culture. From the limited pharmacological analysis of cloned muscarinic receptors that has been completed, it is clear that the muscarinic receptors can be distinguished by a series of selective antagonists when they are expressed in the same cellular environment. For example, the pharmacological profiles of the receptors with respect to some of the most commonly used selective drugs are: pirenzepine (m1 > m3, m4, m5 > m2), AF-DX 116 (m2 > m4 > m3 > m1 > m5), and hexahydrosiladifenidol (m3 > m1, m4, m5 > m2) (Buckley et al., 1989).

Coupling of Cloned Muscarinic Receptors to G Proteins and Physiological Responses

The ability of muscarinic receptors to selectively activate different second messenger systems has been investigated by the transformation of cell lines with the individual muscarinic receptor cDNAs. Stimulation of m1 and m3 receptors expressed by A9 L (Brann et al., 1988b) and HEK (Peralta et al., 1988) cells increases PI metabolism. In A9 L cells, their stimulation also leads to increases in cAMP levels and arachidonic acid release (Brann et al., 1988b; Novotny and Brann, 1989; Conklin et al., 1988). Stimulation of m2 and m4 receptors leads to a decrease in cAMP levels (Brann et al., 1988b; Novotny and Brann, 1989) and, at high receptor levels, also to a weak stimulation of PI metabolism (Peralta et al., 1988). When m3 and m5 receptors are expressed by CHO cells they also increase PI metabolism and cAMP levels (Jones et al., 1989). Responses to m1, m3, and m5 receptor stimulation are usually PTX insensitive, while responses mediated by m2 and m4 receptors are PTX sensitive. The time courses of the m1 receptor–induced cAMP, arachidonic acid, and PI responses in A9 L cells are illustrated in Fig. 1.

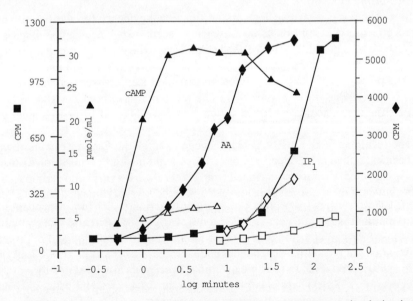

Figure 1. Time courses of the second messenger responses to m1 receptor stimulation in A9 L cells. m1 receptors were stimulated with 100 μM carbachol in the absence (*filled symbols*) and presence (*open symbols*) of 1 μM atropine. *AA*, arachidonic acid; *IP$_1$*, inositol monophosphate.

The first electrophysiological studies of muscarinic receptors involving molecular genetic techniques used injections of mRNA isolated from various tissues into *Xenopus* oocytes. mRNA obtained from rat brain directed synthesis of muscarinic receptors that activated a depolarizing chloride conductance (Sugiyama et al., 1985), while injection of mRNA derived from rat small intestine resulted in muscarinic activation of a variety of depolarizing and hyperpolarizing responses (Aoshima et al., 1987). Using cloned receptors, Fukuda et al. (1987), demonstrated that after injection of m1 mRNA into *Xenopus* oocytes, acetylcholine activated a calcium-dependent chloride conductance, similar to that induced by crude rat brain mRNA injections. Injections of m2 mRNA, however, resulted in activation of a nonselective cation conductance. m2

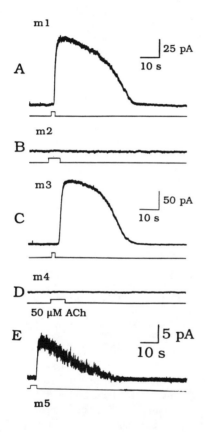

Figure 2. Electrophysiological responses to m1–m5 receptor stimulation. m1–m4 were expressed in A9 L cells and m5 in CHO cells. Cells were voltage clamped at −50 mV using whole-cell patch-clamp recording. Top traces show current responses, lower traces show application of ACh (50 mM) from a pressure ejection micropipette.

receptors also weakly activated a calcium-dependent chloride conductance. In A9 L and NG108-15 cells, m1 receptors predominantly activated calcium-dependent potassium conductances (Fukuda et al., 1988; Jones et al., 1988a), but also weakly activated a calcium-dependent chloride conductance (Jones et al., 1988a). m3 and m5 receptors also activated these calcium-dependent conductances (Fukuda et al., 1988; Jones et al., 1988a; 1989). In NG108-15 cells the m1 and m3 receptors also inhibited the m-current (Fukuda et al; 1988). There was no m-current in A9 L or CHO cells. m2 and m4 receptors produced no electrically detectable responses in A9 L, CHO, and NG108-15 cells (Fukuda et al., 1988; Jones et al., 1988b). Electrophysiological responses of the m1–m5 receptors expressed by A9 L and CHO cells are illustrated in Fig. 2.

As was the case for PI and cAMP metabolism, the m1 and m3 receptor-induced calcium-dependent conductance was PTX insensitive (Jones et al., 1990). Incorporation of IP$_3$ in the patch pipette or artificially raising the intracellular calcium concentration, mimicked the muscarinic activation of the potassium current responses in A9 L cells (Jones et al., 1990), leading to the suggestion that activation of

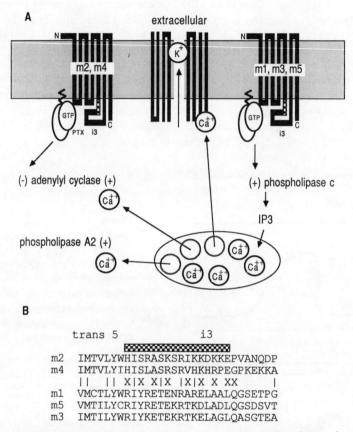

Figure 3. (*A*) Diagramatic representation of m1–m5 receptor-mediated second messenger and electrophysiological responses. The mechanism by which muscarinic receptors mediate the m-current is not illustrated, as its mechanism remains controversial. Checked boxes indicate the region of the third cytoplasmic (i3) loop that has been implicated as being important in second messenger coupling. (*B*) Sequences of the region of the i3 loops that have been implicated in confering specificity of coupling muscarinic receptors to second messengers. *I* indicates the positions at which all muscarinic receptors are either identical or at which only conservative substitutions occur. *X* indicates the positions at which residues are conserved within the two functional classes of receptors (even vs. odd).

phospholipase C by m1, m3, and m5 results in IP$_3$ formation, which releases calcium from intracellular stores, which in turn activates the channels. An increase in intracellular calcium was demonstrated using the fluorescent calcium indicator dye fura-2, on stimulation of m1 and m3 receptors (Neher et al., 1988; Jones et al., 1990) providing more evidence for this mechanism. A summary of the molecular mechanisms of the responses mediated by m1–m5 receptors in A9 L and CHO cells is illustrated in Fig. 3 A.

Structural Determinants of Coupling Selectivity

Various studies have investigated the structural basis of the functional diversity and G protein coupling selectivity of muscarinic receptor subtypes. As G proteins are located at the cytoplasmic surface of the plasma membrane (Gilman, 1987), the putative intracellular domains of the muscarinic receptors are the most likely candidates for conferring coupling selectivity. A recent report suggested that the third cytoplasmic loop (i3) is a major determinant of the functional heterogeneity of muscarinic receptors (Kubo et al., 1988). Expression of chimeric m1/m2 receptors in *Xenopus* oocytes showed that only those constructs that had the m1 sequence in their i3 domain induced the calcium-dependent inward currents characteristic of m1 stimulation. On the other hand, a chimeric receptor composed of the i3 loop from m2 and the core of the m1 receptor mediated the electrophysiological effects typical for m2 receptor activation.

In a related study, chimeric m2/m3 receptors were transiently expressed in COS-7 cells (Wess et al., 1989). Wild type m2 only weakly stimulated PI turnover, while m3 strongly coupled to this effector system. Replacement of the i3 domain of m2 by the corresponding m3 sequence led to a receptor chimera which, similar to m3, induced a strong PI response with a carbachol EC_{50} comparable to that of m3. Since it has been shown that IP_3, a major product of PI hydrolysis mimics the current responses induced by m1 and m3 (Jones et al., 1990), the i3 domain of these receptors specifies both the biochemical and electrophysiological responses.

Analysis of the sequences of the i3 loops of the five muscarinic receptor subtypes indicates that the NH_2-terminal portion of the i3 loop is highly conserved among m1, m3, and m5. The sequences of m2 and m4 are also similar to each other in this region, but differ substantially from m1, m3, and m5 (see Fig. 3 B). To test whether this domain is responsible for defining coupling selectivity, chimeric m2/m3 receptors were created in which this segment has been exchanged between m2 and m3 (Wess et al., 1989). Introduction of 17 amino acids of m3 sequence from the NH_2-terminal portion of i3 into m2 yielded a chimeric receptor that mediated a pronounced stimulation of PI metabolism comparable in magnitude to that of m3. However, the carbachol EC_{50} for inducing this response was about 20-fold higher for this chimeric receptor than for wild type m3. As radioligand binding data suggest that this reduction in potency is not due to a decrease in carbachol affinity, it appears that other as yet unidentified domains of the i3 loop (i.e., the COOH-terminal portion) are involved in conferring full coupling efficiency (Wess et al., 1989). Conversely, deletion mutants of the m1 muscarinic receptor, which lack the majority of the i3 loop (which is not adjacent to the transmembrane domains), are able to potently stimulate PI metabolism (Shapiro and Nathanson, 1989). Together, these observations suggest that the regions of the i3 loop proximal to the transmembrane domains are critical in defining second messenger coupling of muscarinic receptors. These relationships are summarized in Fig. 3, *A* and *B*.

Since work with chimeric $\alpha2/\beta2$ adrenergic receptors has similarly suggested an important role for the i3 loop in coupling selectivity (Kobilka et al., 1988), and since deletion mutants of the β2 receptor, which lack the NH_2 or COOH-terminal sequences of the i3 loop, are unable to stimulate adenylyl cyclase (Strader et al., 1987), these domains may determine the coupling selectivity of all G protein-coupled receptors.

References

Aoshima, H., H. Iio, M. Anan, H. Ishii, and S. Kobayashi. 1987. Induction of muscarinic acetylcholine, serotonin and substance P receptors in *Xenopus* oocytes injected with mRNA prepared from the small intestine of rats. *Molecular Brain Research.* 2:15–20.

Baumgold, J. 1988. Comparison of the coupling and pharmacological characteristics of muscarinic receptors from SK-N-SH human neuroblastoma cells with those from NG108-15 cells. *Life Sciences.* 45:14–118.

Bonner, T. I., N. J. Buckley, A. C. Young, and M. R. Brann. 1987. Identification of a family of muscarinic acetylcholine receptor genes. *Science.* 237:527–532.

Bonner, T. I., A. Young, M. R. Brann, and N. J. Buckley. 1988. Cloning and expression of the human and rat m5 muscarinic receptor genes. *Neuron.* 1:430–410.

Brann, M. R., N. J. Buckley, and T. I. Bonner. 1988a. The striatum and cerebral cortex express different muscarinic receptor mRNAs. *FEBS Letters.* 230:90–94.

Brann, M. R., B. R. Conklin, N. M Dean, R. M. Collins, T. I. Bonner, N. J. Buckley. 1988b. Cloned muscarinic receptors couple to different G-proteins and second messengers. *Society of Neuroscience Abstracts.* 14:600. (Abstr.)

Buckley, N. J., T. I. Bonner, and M. R. Brann. 1988. Localization of a family of muscarinic receptor mRNAs in rat brain. *Journal of Neuroscience.* 8:4646–4652.

Buckley, N. J., T. I. Bonner, C. M. Buckley, and M. R. Brann. 1989. Antagonist binding properties of five cloned muscarinic receptors expressed in CHO-K1 cells. *Molecular Pharmacology.* 35:469–476.

Clapp, L. H., M. B. Vivaudou, J. V. Walsh, and J. J. Singer. 1987. Acetylcholine increases voltage-activated Ca^{2+} current in freshly dissociated smooth muscle cells. *Proceedings of the National Academy of Sciences.* 84:2092–2096.

Conklin, B. R., M. R. Brann, N. J. Buckley, A. Ma, T. I. Bonner, and J. Axelrod. 1988. Stimulation of arachidonic acid release and inhibition of mitogenesis by cloned genes for muscarinic receptor subtypes stably expressed in A9 L cells. *Proceedings of the National Academy of Sciences.* 85:8698–8702.

Constanti, A., and J. A. Sim. 1987. Muscarinic receptors mediating suppression of the M-current in guinea-pig olfactory cortex neurones may be of the M2-subtype. *British Journal of Pharmacology.* 90:3–5.

Egan, T. M., and R. A. North. 1986. Acetylcholine hyperpolarizes central neurones by acting on an M2 muscarinic receptor. *Nature.* 319:405–407.

Evans, M. G., and A. Marty. 1986. Potentiation of muscarinic and α-adrenergic responses by an analogue of guanosine 5'-triphosphate. *Proceedings of the National Academy of Sciences.* 83:4099–4103.

Fukuda, K., H. Higashida, T. Kubo, A. Maeda, I. Akiba, H. Bujo, M. Mishina, and S. Numa. 1988. Selective coupling with K^+ currents of muscarinic ACh receptor subtypes in NG108-15 cells. *Nature.* 335:355–358.

Fukuda, K., T. Kubo, I. Akiba, A. Maeda, M. Mishina, and S. Numa. 1987. Molecular distinction between muscarinic acetylcholine receptor subtypes. *Nature.* 327:623–625.

Gähwiler, B. H., and D. A. Brown. 1987. Muscarine affects calcium currents in rat hippocampal pyramidal cells in vitro. *Neuroscience Letters.* 76:301–306.

Gallacher, D. V., and A. P. Morris. 1987. The receptor-regulated calcium influx in mouse submandibular acinar cells is sodium dependent: a patch-clamp study. *Journal of Physiology.* 384:119–130.

Galligan, J. J., R. A. North, and T. Tokimasa. 1988. Muscarinic agonists and potassium currents in the guinea-pig myenteric neurones. *British Journal of Pharmacology.* 219:213–220.

Gilman, A. 1987. G-proteins: transducers of receptor-generated signals. *Annual Review of Biochemistry.* 56:615–649.

Haga, K., and T. Haga. 1985. Purification of the muscarinic acetylcholine receptor from procine brain. *Journal of Biological Chemistry.* 260:7927–7935.

Harden, T. K., L. I. Tanner, M. W. Martin, N. Nakahata, A. R. Hugher, J. R. Hepler, T. Evans, S. B. Masters, and J. H. Brown. 1986. Characteristics of two biochemical responses to stimulation of muscarinic cholinergic receptors. *Trends in Pharmacological Sciences, Supplement.* 7:14–18.

Jones, S. V. P., J. L. Barker, T. I. Bonner, N. J. Buckley, and M. R. Brann. 1988a. Electrophysiological characterization of the cloned m1 muscarinic receptor expressed in A9 L cells. *Proceedings of the National Academy of Sciences.* 85:4056–4060.

Jones, S. V. P., J. L. Barker, N. J. Buckley, T. I. Bonner, R. C. Collins, and M. R. Brann. 1988b. Cloned muscarinic receptor subtypes expressed in A9 L cells differ in their coupling to electrical responses. *Molecular Pharmacology.* 34:421–426.

Jones, S. V. P., J. L. Barker, M. B. Goodman, and M. R. Brann. 1990. Inositol trisphosphate mediates cloned muscarinic receptor-activated conductances in transfected mouse fibroblast A9 L cells. *Journal of Physiology.* 421:499–519.

Jones, S. V. P., T. J. Murphy, and M. R. Brann. 1989. Physiological comparison of cloned muscarinic receptor subtypes expressed in CHO cells. *Trends in Pharmacological Sciences, Supplement.* 10:116–117 (Abstr.)

Kirsch, G. E., A. Yatani, J. Condina, L. Birnbaumer, and A. M. Brown. 1988. α-subunit of G_K activates atrial K^+ channels of chick, rat and guinea-pig. *American Journal of Physiology.* 254:H1200–H1205.

Kobilka, B. K., T. S. Kobilka, K. Daniel, J. W. Regan, M. G. Caron, and R. J. Lefkowitz. 1988. Chimeric $\alpha2$-, $\beta2$-adrenergic receptors: delineation of domains involved in effector coupling and ligand binding specificity. *Science.* 240:1310–1316.

Kubo, T., H. Bujo, I. Akiba, J. Nakai, M. Mishina, and S. Numa. 1988. Location of a region of the muscarinic acetylcholine receptor involved in selective effector coupling. *FEBS Letters.* 241:119–125.

Kubo, T., K. Fukada, A. Mikami, A. Maeda, H. Takahashi, M. Mishina, K. Haga, T. Haga, A. Ichiyama, K. Kangawa, M. Kojima, H. Matsuo, T. Hirose, and S. Numa. 1986. Cloning, sequencing and expression of complementary DNA encoding the muscarinic ACh receptor. *Nature.* 323:411–416.

Madison, D. V., B. Lancaster, and R. A. Nicoll. 1987. Voltage-clamp analysis of cholinergic action in the hippocampus. *Journal of Neuroscience.* 7:733–741.

Maeda, A., T. Kubo, M. Mishina, and S. Numa. 1988. Tissue distribution of mRNAs encoding muscarinic acetylcholine receptor subtypes. *FEBS Letters.* 239:339–342.

McCormick, D. A., and H. C. Pape. 1988. Acetylcholine inhibits identified interneurons in the cat lateral geniculate nucleus. *Nature.* 334:246–248.

McCormick, D. A., and D. A. Prince. 1986. Acetylcholine induces burst firing in thalamic reticular neurones by activating a potassium conductance. *Nature.* 44:402–404.

McCormick, D. A. and D. A. Prince. 1987. Mechanisms of action of acetylcholine in the guinea-pig cerebral cortex in vitro. *Journal of Physiology.* 375:169–191.

Müller, W., and U. Misgeld. 1986. Slow cholinergic excitation of guinea pig hippocampal neurons is mediated by two muscarinic receptor subtypes. *Neuroscience Letters.* 67:107–112.

Mutschler, E., G. Lambrecht, and J. Wess. 1988. Selective muscarinic agonists and antagonists: a challenge for the medicinal chemist. *In* Xth International Symposium on Medicinal Chemistry. Budapest, Hungary.

Nathanson, N. M. 1987. Molecular properties of the muscarinic acetylcholine receptor. *Annual Review of Neuroscience.* 10:195–236.

Neher, E., A. Marty, K. Fukuda, T. Kubo, and S. Numa. 1988. Intracellular calcium release mediated by two muscarinic receptor subtypes. *FEBS Letters.* 240:88–94.

Novotny, E., and M. R. Brann. 1989. Agonist pharmacology of cloned muscarinic receptors. *Trends in Pharmacological Sciences, Supplement.* 10:116. (Abstr.)

Peralta, E. G., A. Ashkenazi, J. W. Winslow, J. Ramachandran, and D. C. Capon. 1988. Differential regulation of PI hydrolysis and adenylyl cyclase by muscarinic receptor subtypes. *Nature.* 334:434–437.

Peralta, E. G., A. Ashkenazi, J. W. Winslow, D. H. Smith, J. Ramachandran, and D. J. Capon. 1987a. Distinct primary structures, ligand-binding properties and tissue-specific expression of four human muscarinic acetylcholine receptors. *EMBO Journal.* 6:3923–3929.

Peralta, E. G., J. W. Winslow, G. L. Peterson, D. H. Smith. A. Ashkenazi, J. Ramachandran, M. J. Schimerlik, and D. J. Capon. 1987b. Primary structure and biochemical properties of an M2 muscarinic receptor. *Science.* 236:600–605.

Peterson, G. L., G. S. Herron, M. Yamaki, D. S. Fullerton, and M. I. Schimerlik. 1984. Purification of the muscarinic acetylcholine receptor from porcine atria. *Proceedings of the National Academy of Sciences.* 81:4993–4997.

Pfaffinger, P. J., J. M. Martin, D. D. Hunter, N. M. Nathanson, and B. Hille. 1985. GTP-binding proteins couple cardiac muscarinic receptors to a K channel. *Nature.* 317:536–538.

Sakmann, B., A. Noma, and W. Trautwein. 1983. Acetylcholine activation of single muscarinic K^+ channels in isolated pacemaker cells of mammalian heart. *Nature.* 303:250–253.

Shapiro, R. A., and N. M. Nathanson. 1989. Deletion analysis of the mouse m1 muscarinic acetylcholine receptor: effects on phosphoinositide metabolism and down-regulation. *Biochemistry.* 28:8946–8950.

Simmons, M. A., and H. C. Hartzell. 1987. A quantitative analysis of the acetylcholine-activated potassium current in single cells from frog atrium. *Pflügers Archiv.* 409:454–461.

Strader, C. D., R. A. F. Dixon, A. H. Cheung, M. R. Candelore, A. D. Blake, and I. S. Sigal. 1987. Mutations that uncouple the β adrenergic receptor from Gs and increase agonist affinity. *Journal of Biological Chemistry.* 262:16439–16443.

Stroud, R. M., and J. Finer-Moore. 1985. Acetylcholine receptor structure, function and evolution. *Annual Review of Cellular Biology.* 1:317–351.

Sugiyama, H., Y. Hisanaga, and C. Hirono. 1985. Induction of muscarinic cholinergic responsiveness in *Xenopus* oocytes by mRNA isolated from rat brain. *Brain Research*. 338:346–350.

Toselli, M., and H. D. Lux. 1989. GTP-binding proteins mediate acetylcholine inhibition of voltage-dependent calcium channels in hippocampal neurons. *Pflügers Archiv*. 413:319–321.

Vivaudou, M. B., L. H. Clapp, J. V. Walsh, and J. J. Singer. 1988. Regulation of one type of Ca^{2+} current in smooth muscle cells by diacylglycerol and acetylcholine. *FASEB Journal*. 2:2497–2504.

Wanke, E., A. Ferroni, A. Malgaroli, A. Ambrosini, T. Pozzan, and J. Meldolesi. 1987. Activation of a muscarinic receptor selectively inhibits a rapidly inactivated Ca^{2+} current in rat sympathetic neurons. *Proceedings of the National Academy of Sciences*. 84:4313–4317.

Weiner, D. M., and M. R. Brann. 1989. Distribution of m1–m5 muscarinic receptor mRNAs in rat brain. *Trends in Pharmacological Sciences, Supplement*. 10:115. (Abstr.)

Wess, J., M. R. Brann, and T. I. Bonner. 1989. Identification of a small intracellular region of the muscarinic m3 receptor as a determinant of selective coupling to PI turnover. *FEBS Letters*. 258:133–136.

Chapter 9

Chimeras of the G_s α Subunit That Had the NH_2- or COOH-Terminal Sequence Substituted for the Corresponding Region of the G_i α Subunit Constitutively Activate Adenylyl Cyclase

Lynn E. Heasley, N. Dhanasekaran, Sunil K. Gupta, Shoji Osawa, and Gary L. Johnson

Division of Basic Sciences, National Jewish Center for Immunology and Respiratory Medicine, Denver, Colorado 80206; and the Department of Pharmacology, University of Colorado Medical School, Denver, Colorado 80262

Introduction

G_s and G_i stimulate and inhibit adenylyl cyclase, respectively (Gilman, 1984, 1987). The α_s and α_i polypeptides are ~65% homologous in primary sequence (Jones and Reed, 1987), share common $\beta\gamma$ subunits (Cerione et al., 1986), and selectively couple to different receptors (Wessling-Resnick et al., 1987). Also intrinsic to the tertiary structure of α_s and α_i are the GDP/GTP-binding domains and GTPase activity (Gilman, 1984, 1987; Holbrook and Kim, 1989). The GDP/GTP-binding and GTPase domains are composed of four regions in the primary sequence of the α subunit polypeptide (Holbrook and Kim, 1989) and are homologous to the corresponding domains in p21ras (Holbrook and Kim, 1989) and EFTμ (Jurnak, 1985).

Previously, Masters et al. (1988) demonstrated that the COOH-terminal 40% of the α_s polypeptide encoded the necessary structural information that is critical to both receptor selectivity and adenylyl cyclase stimulation. In the present work, we have used chimeric cDNAs that had ~10% of either the NH_2 or COOH terminus of the α_s polypeptide substituted for the corresponding region of α_{i2}. Substitution of the α_{i2} sequences at the ends of the α_s polypeptide introduces multiple mutations into these regions of the α_s subunit. Expression of the NH_2- or COOH-terminal chimeras in Chinese hamster ovary (CHO K-1) cells resulted in the constitutive activation of adenylyl cyclase. Based on these results, we propose that the NH_2- and COOH-terminal regions of the α_s polypeptide function as intramolecular modulators of α_s activity. The mutant phenotypes are used to develop a model of our understanding of the functional regulatory domains within the α_s polypeptide.

Methods

Plasmids and Transfection Protocol

pCW1 is a pUC13 derivative containing the SV40 enhancer, replication origin and early promoter and the SV40 splicing and polyadenylation sequences (Woon et al., 1988). A Hind III cloning site allows the expression of cDNAs under the control of the SV40 early promoter. The plasmid also carries the TN5 gene for G418 resistance under the control of the MoLTR (Woon et al., 1989). CHO K-1 cells were transfected by electroporation of 10^7 cells using 10 μg plasmid DNA (Gene Pulser: 25 μFD, 1 kV; Bio-Rad Laboratories, Richmond, CA). G418-resistant CHO cell colonies were isolated and screened for chimeric α_i/α_s expression by both Northern and immunoblotting analysis (Woon et al., 1988, 1989).

Other Assays

Cyclic AMP was measured using a radioimmunoassay kit from Amersham Corp. (Arlington Heights, IL) and cAMP-dependent protein kinase was assayed as previously described (Woon et al., 1988, 1989). Adenylyl cyclase was measured by conversion of $[\alpha\text{-}^{32}P]ATP$ to $[^{32}P]cAMP$ (Woon et al., 1988).

Results and Discussion

Three chimeras of α_s and α_{i2} were constructed that result in significant mutation of amino acid sequences within the α_s polypeptide (Fig. 1). The $\alpha_{i(54)/s}$ chimera has the first 61 amino acids of α_s substituted for the first 54 residues of α_{i2}. There are seven additional amino acids in α_s compared with α_{i2} within this region of the NH_2 terminus and 16 of the first 34 α_{i2} residues are nonconserved relative to the α_s sequences. The result is a net charge within nonconserved residues of +6 for α_s and +3 for α_{i2}. The last 20 α_{i2} amino acids (Lys35–Lys54) in the chimera are identical or highly conserved when compared with the α_s sequence.

The second chimera, $\alpha_{s/i(38)}$, encodes the first 356 amino acids of α_s and has the last 38 amino acids of the α_s polypeptide substituted for the COOH-terminal 36 amino acids of α_{i2} (Woon et al., 1988). Within the substituted COOH-terminal sequence there are 17 nonconserved residues and a net charge of +2 for α_s and −1 for α_{i2}. The

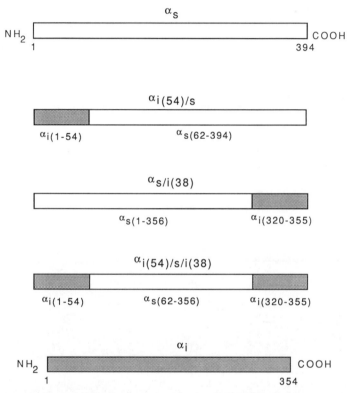

Figure 1. Schematic diagram of the α_i/α_s chimeras. The α_s polypeptide is 394 amino acids while α_{i2} is encoded by 354 residues. The $\alpha_{i(54)/s}$ chimera is encoded by the first 54 residues of α_{i2} and amino acids 62–394 of α_s. The $\alpha_{s/i(38)}$ chimera is encoded by the first 356 residues of α_s and amino acids 320–355 of α_{i2}. The $\alpha_{i(54)/s/i(38)}$ chimera is the combination of the NH$_2$-terminal $\alpha_{i(54)/s}$ and COOH-terminal $\alpha_{s/i(38)}$ chimeras.

final chimera, $\alpha_{i(54)/s/i(38)}$ is a combination of the NH$_2$-terminal $\alpha_{i(54)/s}$ and $\alpha_{s/i(38)}$ constructs resulting in mutation of both ends of the α_s polypeptide.

Stable Expression of α_s/α_i Chimeras in CHO Cells

Levels of cAMP and the activity of cAMP-dependent protein kinase in CHO clones stably expressing the chimeric polypeptides were compared with clones transfected with the wild-type α_s and α_i cDNAs (Table I). Stable expression of $\alpha_{i(54)/s}$, $\alpha_{s/i(38)}$, or the double chimera $\alpha_{i(54)/s/i(38)}$ leads to a two- to threefold increase in intracellular cAMP levels, resulting in the constitutive activation (three- to fourfold) of cAMP-dependent protein kinase. In contrast, expression of the wild-type α_s or α_{i2} polypeptides using the pCW1 expression plasmid had little effect on the cAMP levels or the cAMP-dependent protein kinase activity in CHO cells.

In the presence of the phosphodiesterase inhibitor, methyl isobutylxanthine (IBMX), the constitutive increase in cAMP synthesis between the clones expressing the chimeric constructs and the wild-type gene products was even more pronounced. The significantly enhanced cAMP levels in the presence of IBMX is indicative of

TABLE I
Cyclic AMP Levels and Protein Kinase Activity in Stable CHO Transfectants

Clone	cAMP		cAMP-dependent protein kinase		
	−IBMX	+IBMX	Basal	+cAMP	Activity ratio
	pmol/mg protein		*pmol/min per mg protein*		
Wild-type CHO	4.6	22.4	231.1	2,527	0.09
Neo−4	4.6	15.4	215.2	2,873	0.07
α_s-14	3.8	32.6	176.3	1,916	0.09
α_i-2	3.9	30.4	139.7	1,710	0.08
$\alpha_{i(54)/s}$-10	10.4	52.5	512.3	2,754	0.19
$\alpha_{i(54)/s}$-12	14.8	62.7	510.5	2,391	0.21
$\alpha_{i(54)/s}$-13	11.9	71.7	370.6	2,599	0.14
$\alpha_{i(54)/s}$-15	14.9	89.8	670.8	2,506	0.27
$\alpha_{s/i(38)}$-2	8.6	231	440.0	1,600	0.28
$\alpha_{s/i(38)}$-3	8.5	225	440.0	1,490	0.30
$\alpha_{s/i(38)}$-9	12.7	405	640.0	2,130	0.30
$\alpha_{i(54)/s/i(38)}$-3	13.6	75.6	439.8	2,406	0.18
$\alpha_{i(54)/s/i(38)}$-11	12.7	83.7	503.4	2,662	0.19
$\alpha_{i(54)/s/i(38)}$-13	8.4	72.7	318.7	2,382	0.13
$\alpha_{i(54)/s/i(38)}$-24	9.5	94.0	917.8	3,242	0.28

CHO cells were transfected by electroporation using the appropriate pCW1 construct, and G418-resistant clones expressing the α subunit transcripts were identified by Northern analysis. Independent clones were incubated (10 min, room temperature) in the absence or presence of 500 μM IBMX. Cellular cAMP was extracted with 1 ml of 60% ethanol and assayed with a radioimmunoassay kit from Amersham Corp. cAMP-dependent protein kinase was assayed as described previously (Woon et al., 1988, 1989); the activity ratio is the basal cAMP-dependent kinase activity divided by the maximal activity determined in the presence of 3 μM cAMP.

increased cAMP phosphodiesterase activity in the $\alpha_{i(54)/s}$-, $\alpha_{s/i(38)}$-, and $\alpha_{i(54)/s/i(38)}$-expressing clones. This phenomenon has been observed in other cell types where cAMP analogues or hormones have chronically activated cAMP-dependent protein kinase and resulted in the induction of phosphodiesterase activity (Insel et al., 1975). It is significant, therefore, that in the presence of IBMX, neither the α_s- or α_i-expressing clones show a marked increase in intracellular cAMP compared with untransfected CHO cells or cells transfected with the pCW1 expression vector (Table I). Thus, the constitutive activation of adenylyl cyclase activity in CHO cells results in compensatory increases in cAMP phosphodiesterase activity presumably as a means to blunt the elevated intracellular cAMP levels. In addition to the increase in cAMP phosphodiesterase activity observed with $\alpha_{i(54)/s}$-expressing clones, there was a modest 1.1–1.3-fold increase in β subunit expression as determined by immunoblotting and densitometry (not shown). CHO clones transfected with wild-type α_s or α_i showed no change in β subunit expression. Nonetheless, even with the enhanced cAMP phosphodiesterase activity and modest increase in β subunit expression, the CHO cells expressing the chimeric polypeptides exhibited constitutively activated cAMP-dependent protein kinase activity. The α subunit chimera–expressing clones have shown the constitutively elevated cAMP levels and kinase activation for >100 cell doublings. These results indicate that the constitutive elevation of cAMP levels and kinase activation are unique properties of the $\alpha_{i(54)/s}$, $\alpha_{s/i(38)}$, and $\alpha_{i(54)/s/i(38)}$ chimeras, which behave as dominant activated α_s molecules in stable CHO cell transfectants.

TABLE II
Adenylyl Cyclase Regulation in Stable CHO Transfectants

Clone	Lag time to achieve GTPγS-stimulated V_{max}	GTPγS-stimulated V_{max}	GTPase k_{off}
	min	*pmol/min per mg protein*	*min^{-1}*
Wild-type	5–6	100	5–6
α_s	5–6	100	5–6
$\alpha_{i(54)/s}$	1.5–2	100	5–6
$\alpha_{s/i(38)}$	1.5–2	350	5–6
$\alpha_{i(54)/s/i(38)}$	1.5–2	350	5–6
α_sLeu227	5–6	150	0.05–0.07

Membranes from the designated CHO cell clones were prepared and solubilized in 1% sodium cholate, which denatures catalytic adenylyl cyclase. Equivalent amounts of protein in extracts containing the G proteins were reconstituted into *cyc*-S49 membranes and assayed for GTPγS-regulable adenylyl cyclase activity (Woon et al., 1988). GTPase k_{off} measurements were performed with cholate extracts reconstituted with *cyc*-membranes by determining the rate of propranolol inhibition of isoproterenol-stimulated adenylyl cyclase activity in the presence of GTP (Masters et al., 1989).

Adenylyl Cyclase Regulation in CHO Cell Transfectants

Membranes from CHO clones expressing $\alpha_{i(54)/s}$, $\alpha_{s/i(38)}$, or $\alpha_{i(54)/s/i(38)}$ chimeras exhibit an altered GTPγS and fluoride-regulated adenylyl cyclase relative to CHO clones expressing wild-type α_s (Table II). Furthermore, the phenotype of clones expressing the NH$_2$-terminal $\alpha_{i(54)/s}$ chimera is similar, but distinct, from the properties of the COOH-terminal $\alpha_{s/i(38)}$ chimera. The $\alpha_{i(54)/s}$ and $\alpha_{s/i(38)}$ phenotype is manifested in adenylyl cyclase assays as a significant decrease in the time required to achieve maximal adenylyl cyclase activation by the hydrolysis-resistant GTP analogue, GTPγS. This decreased lag in achieving maximal GTPγS activation was observed in multiple independent $\alpha_{i(54)/s}$-, $\alpha_{s/i(38)}$-, and $\alpha_{i(54)/s/i(38)}$-expressing CHO clones. The diminished lag time associated with the chimeras, but not wild-type α_s or α_i, is most simply explained by an accelerated rate of GDP dissociation, which would allow a faster GTPγS binding and thus, adenylyl cyclase activation.

In addition to the diminished lag time for maximal GTPγS activation, the COOH-terminal chimeras (i.e., $\alpha_{s/i(38)}$ and $\alpha_{i(54)/s/i(38)}$), but not the NH$_2$-terminal chimera ($\alpha_{i(54)/s}$), demonstrate a marked three- to fourfold increase in the V_{max} of adenylyl cyclase stimulation by either fluoride or GTPγS (Table II). The increased adenylyl cyclase V_{max} observed in the $\alpha_{s/i(38)}$- and $\alpha_{i(54)/s/i(38)}$-expressing clones cannot be explained by an enhanced GDP dissociation rate, but rather, must result from altered intrinsic regulation of α_s activation of adenylyl cyclase. Thus, mutation at the COOH-terminus of α_s not only results in a similar decreased lag in GTPγS activation of adenylyl cyclase, as was seen with NH$_2$-terminal mutation of α_s, but also markedly enhances maximal stimulation of cAMP synthesis, indicating that the NH$_2$- and COOH-terminal ends of α_s have overlapping but distinguishable regulatory functions in controlling α_s activity.

Table II also summarizes a kinetic analysis of adenylyl cyclase activity in the various CHO clones which indicates that the GTPase activity of the chimeric polypeptides was similar to the wild-type α_s. In this analysis, the time course of propranolol-induced inhibition of isoproterenol-stimulated adenylyl cyclase activation was measured, allowing the estimation of the GTPase turn-off rate constant (k_{off}). Evaluation of the k_{off} using this indirect method has been shown to be a reliable measure of the k_{cat} for the intrinsic GTPase reaction of α_s and α_s mutants (Graziano and Gilman, 1989; Masters et al., 1989). The wild-type α_s polypeptides exhibit a k_{off} of ~5–6 min^{-1}. A similar k_{off} was estimated in clones expressing $\alpha_{i(54)/s}$, $\alpha_{s/i(38)}$, and $\alpha_{i(54)/s/i(38)}$, indicating that the GTPase activity of the chimeric polypeptides is not affected. In contrast, the α_sLeu227 mutation, whose GTPase k_{off} is ~0.05 min^{-1} (Graziano and Gilman, 1989; Masters et al., 1989), inhibits the rate of GTP hydrolysis >95%.

Predicted tertiary structural properties of the α_s and α_i subunits, drawn from analogy with the p21ras polypeptide, provide a basis for understanding the NH$_2$- and COOH-terminal mutations (i.e., $\alpha_{i(54)/s}$ and $\alpha_{s/i(38)}$) and the mechanism by which they constitutively activate the G protein, G$_s$. The tertiary structure of p21ras, deduced from x-ray crystallography (Pai et al., 1989), roughly defines the molecule in two "halves" or domains with the guanine nucleotide–binding site in proximity to the cleft between the halves. The crystal structure also shows the NH$_2$- and COOH-terminal ends of the p21ras molecule in close proximity to each other. A similar model for a prototypical α subunit of a G protein has been recently proposed (Holbrook and Kim, 1989) and there is physical evidence for G protein α subunit NH$_2$- and COOH-terminal ends to be in close proximity to one another in the tertiary structure of the polypeptide (Hingorani and Ho, 1988). We propose that the structure of the α_s polypeptide is similar to p21ras and is roughly defined in two "halves" with the NH$_2$ and COOH termini in close proximity. The GTP-binding site would lie in proximity to the groove between the NH$_2$- and COOH-terminal halves similar to that for p21ras. Previous studies indicate that the COOH-terminal 40% of the α_s polypeptide encodes the necessary information for both adenylyl cyclase stimulation and selectivity for receptor coupling (Masters et al., 1988). Our studies demonstrate that the regulatory function of the NH$_2$-terminal moiety of α_s is clearly independent of the intrinsic GTPase activity and instead represents an intramolecular attenuator of GDP dissociation and α_s activation. Thus, we propose that the termini of the α_s polypeptide function as modulators of the two α_s domains to control both activation and attenuation. When this control is lost because of mutation, as in the $\alpha_{i(54)/s}$ and $\alpha_{s/i(38)}$ chimeras, the phenotype is a constitutively activated α_s molecule. The similarly diminished lag time for activation of adenylyl cyclase by $\alpha_{i(54)/s}$, $\alpha_{s/i(38)}$, and $\alpha_{i(54)/s/i(38)}$ chimeras indicates that the NH_2 and COOH termini of the α subunit polypeptide have overlapping functions in the control of the attenuator domain. The marked increase in V_{max} for adenylyl cyclase stimulation by GTPγS observed only with the COOH-terminal mutants indicates that this region of the molecule directly influences cyclase activation by the $\alpha_s \cdot$ GTPγS complex.

A prediction, if this model is correct, is that mutations in the modulator domains (NH_2 and COOH termini) should be additive with mutations that inhibit the GTPase turn-off activity (α_sLeu227). In fact, introduction of the Leu227 mutation into $\alpha_{i(54)/s}$ resulted in an α subunit which elevated cAMP in different cell types after gene transfection to levels that were additive of the cAMP levels produced by the two individual mutations (manuscript in preparation). This finding provides a structural basis to construct strong, dominant, constitutively active G protein α subunits that have lost both GTPase shut-off and attenuation of GDP dissociation. The influence of dominant constitutively active G protein α subunit polypeptides on cell function can now be systematically addressed.

Acknowledgments

This work was supported by National Institutes of Health grants GM-30324 and DK-37871.

References

Cerione, R. A., C. Staniszewski, P. Gierschik, J. Codina, R. L. Somers, L. Birnbaumer, A. M. Spiegel, M. G. Caron, and R. J. Lefkowitz. 1986. Mechanism of guanine nucleotide regulatory protein-mediated inhibition of adenylate cyclase. *Journal of Biological Chemistry*. 261:9514–9520.

Gilman, A. G. 1984. G proteins and dual control of adenylate cyclase. *Cell*. 39:577–579.

Gilman, A. G. 1987. G proteins: transducers of receptor-generated signals. *Annual Review of Biochemistry*. 56:615–649.

Graziano, M. P., and A. G. Gilman. 1989. Synthesis in *E. coli* of GTPase-deficient mutants of $G_s\alpha$. *Journal of Biological Chemistry*. 264:15475–15482.

Hingorani, V. N., and L. K. Ho. 1988. Fluorescent labeling of the signal-transducing G proteins: pertussis toxin-catalyzed etheno-ADP-ribosylation of transducin. *Journal of Biological Chemistry*. 263:19804–19808.

Holbrook, S. R., and S.-H. Kim. 1989. Molecular model of the G protein α subunit based on the crystal structure of the HRAS protein. *Proceedings of the National Academy of Sciences*. 86:1751–1755.

Insel, P. A., H. R. Bourne, P. Coffino, and G. M. Tomkins. 1975. Cyclic AMP dependent protein kinase: pivotal role in regulation of enzyme induction and growth. *Science*. 190:896–898.

Jones, D. T., and R. R. Reed. 1987. Molecular cloning of five GTP-binding protein cDNA species from rat olfactory neuroepithelium. *Journal of Biological Chemistry*. 262:14241–14249.

Jurnak, F. 1985. Structure of the GDP domain of EF-Tμ and location of the amino acids homologous to *ras* oncogene proteins. *Science*. 230:32–36.

Masters, S. B., T. R. Miller, M.-H. Chi, F.-H. Chang, B. Beiderman, N. G. Lopez, and H. R. Bourne. 1989. Mutations in the GTP-binding site of $G_s\alpha$ alter stimulation of adenylyl cyclase. *Journal of Biological Chemistry.* 264:15467–15474.

Masters, S. B., K. A. Sullivan, R. T. Miller, B. Biederman, N. G. Lopez, J. Ramachandran, and H. R. Bourne. 1988. Carboxyl terminal domain of $G_s\alpha$ specifies coupling of receptors to stimulation of adenylyl cyclase. *Science.* 241:448–451.

Pai, E. F., W. Kabsch, U. Krengel, K. C. Holmes, J. John, and A. Wittinghofer. 1989. Structure of the guanine nucleotide binding domain of the Ha-ras oncogene product p21 in the triphosphate conformation. *Nature.* 341:209–214.

Wessling-Resnick, M., D. J. Kelleher, E. R. Weiss, and G. L. Johnson. 1987. Enzymatic model for receptor activation of GTP-binding proteins. *Trends in Biochemical Sciences.* 12:473–477.

Woon, C. W., L. Heasley, S. Osawa, and G. L. Johnson. 1989. Mutation of glycine 49 to valine in the α subunit of G_s results in the constitutive elevation of cyclic AMP synthesis. *Biochemistry.* 28:4547–4551.

Woon, C. W., S. Soparkar, L. Heasley, and G. L. Johnson. 1988. Expression of a $G\alpha_s/G\alpha_i$ chimera that constitutively activates cyclic AMP synthesis. *Journal of Biological Chemistry.* 264:5687–5693.

Chapter 10

G Protein–linked Signal Transduction in Aggregating *Dictyostelium*

Geoffrey S. Pitt, Robert E. Gundersen, Pamela J. Lilly, Maureen
B. Pupillo, Roxanne A. Vaughan, and Peter N. Devreotes

*Department of Biological Chemistry, The Johns Hopkins
University School of Medicine, Baltimore, Maryland 21205*

Introduction

The social amoeba *Dictyostelium* displays G protein–linked signal transduction pathways that are analogous to classical models, such as β-agonist stimulation in the heart and photon stimulation of rhodopsin in the retina, and that reveal complexities of these pathways now being explored in higher eukaryotes. The remarkable degree of similarity in the biochemistry and molecular components of these pathways among *Dictyostelium* and higher eukaryotes emphasizes the conservation of this signal transduction motif. Because *Dictyostelium* offers a biochemically and genetically accessible system, it is a practical model that is being successfully exploited to study G protein–linked signal transduction.

Signal Transduction in Aggregating *Dictyostelium*

A combination of biochemical evidence and the study of mutants has identified at least two G protein–coupled pathways present during the early part of the *Dictyostelium* developmental life cycle (Theibert and Devreotes, 1986; Van Haastert et al., 1986). This developmental program, triggered by nutrient starvation, is marked by the aggregation of single-cell amoebae into a multicellular organism for the purpose of sporulation. Aggregation is coordinated by pulsatile waves of cAMP, secreted by cells at aggregation centers, to which surrounding cells respond both by displaying chemotaxis, mediated through their chemoattractant receptor, and by relaying the signal to cells further from the center by secreting cAMP (Klein et al., 1988). When each peak of a cAMP wave passes, cells adapt to the cAMP signal and temporarily cease both chemotaxis and signaling. This assures the directed movement of the cells toward the aggregation center and prevents them from following the cAMP peak away from the center. Secreted phosphodiesterase (PDE) depletes the extracellular cAMP so that cells can resensitize in preparation for the next cAMP oscillation (Darmon et al., 1978). The aggregation-deficient Frigid A (*fgd A*) (Coukell, 1975) and *synag* (Pupillo et al., 1988) mutations demonstrate that both the chemotactic and signaling responses are mediated by G protein–linked cAMP receptors and help to elucidate the signal transduction pathways.

The current model, as depicted in Fig. 1, postulates that cAMP binds to the chemoattractant receptor to activate a G protein. This G protein interacts with effector(s) such as guanylate cyclase (Van Haastert et al., 1986) to elicit the chemotactic response. Adaptation of this response coincides with ligand-induced phosphorylation of the cAMP receptor (Vaughan and Devreotes, 1988).

In *fgd A* cells, the chemotactic response is blocked; these cells fail to aggregate into the multicellular organism necessary for sporulation. Binding studies with cAMP and Western blotting with receptor-specific antisera illustrate that these cells express cAMP receptors capable of binding cAMP. In membranes prepared from *fgd A* cells, guanine nucleotide regulation of receptor affinity for cAMP, however, is decreased when compared with membranes from wild-type cells. These data, coupled with evidence showing that wild-type cAMP–stimulated responses are absent or decreased in *fgd A* cells, suggested that the *fgd A* mutation blocks interaction between receptors and a G protein (Kesbeke et al., 1988). Cloning of the cDNAs for two G protein α subunits from *Dictyostelium* (Pupillo et al., 1989) allowed the *fgd A* locus to be assigned to the Gα2 gene, indicating the cAMP responses are mediated through the Gα2 protein (Kumagai et al., 1989).

In wild-type cells, cAMP stimulation induces reversible phosphorylation of Gα2 on serine residue(s) (Gundersen and Devreotes, 1990). The kinetics and dose response of this modification parallel the transient activation of many cAMP-induced responses, suggesting that Gα2 phosphorylation may be involved in cAMP-induced activation of intracellular effectors. This first demonstration of ligand-induced G protein modification is an intriguing observation that may prove to be universal in G protein signal transduction systems.

Although Gα2 appears to be required for cAMP stimulation of adenylate cyclase, since this response is blocked in *fgd A* cells, Gα2 is not the G protein that interacts with the catalytic component of adenylate cyclase directly. This is demonstrated by experiments which show that guanine nucleotides can regulate adenylate cyclase in membranes prepared from wild-type cells as well as from cells of *fgd A* alleles that fail to express Gα2 protein (Kesbeke et al., 1988).

The hypothesis that another G protein interacts with adenylate cyclase, and is thus part of the signalling response, also results from the study of *synag* cells, a second class of mutants which fails to aggregate. This class is so named because mutants can

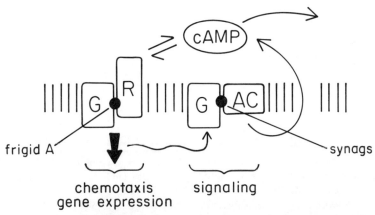

Figure 1. Schematic of the signal transduction pathway in *Dictyostelium*. Extracellular cAMP binds to a transmembrane cAMP receptor, which interacts through a G protein, Gα2, to promote chemotaxis and gene expression. Stimulation by extracellular cAMP also results in activation of adenylate cyclase and secretion of cAMP. The *fgd A* mutations block in vivo cAMP increases, chemotaxis, and guanine nucleotide regulation of receptor affinity. The *fgd A* mutations do not block GTP stimulation of adenylate cyclase in vitro, but cAMP does not enhance GTP stimulation. The *fgd A* locus has been shown to be the Gα2 gene. The *synag* mutations block both GTP stimulation of adenylate cyclase and in vivo cAMP increase, but do not block guanine nucleotide regulation of cAMP receptor affinity or chemotaxis.

be rescued by synergy with wild-type cells or exogenously supplied cAMP. The G protein–linked chemotactic response thus appears intact.

In one set of *synag* mutants adenylate cyclase activity cannot be stimulated by cAMP, suggesting that these mutations block the signalling response. This set can be further subdivided based on biochemical analysis. In membranes prepared from four *synag* mutants (Table I), guanine nucleotides do not regulate adenylate cyclase activity, suggesting that the block in these mutants lies in G protein regulation of adenylate cyclase (Theibert and Devreotes, 1986). When unregulated adenylate cyclase activity, as assayed by the addition of Mn^{++} to membranes, is examined in these four, a subset is defined (*synag 24* and *synag 49*) in which this activity is reduced. This suggests that these mutations lie in the catalytic component of adenylate cyclase.

The deficiency in G protein regulation of adenylate cyclase activity of one mutant, *synag 7* (which possesses Mn^{++}-stimulated adenylate cyclase activity), can be rescued with the addition of a component from the 100,000 g supernatant from lysates of *synag 49* (Theibert and Devreotes, 1986). This soluble factor has been named GRP (GTP-activating reconstituting protein). When supernatants from wild-type and *synag 7* lysates were compared by two-dimensional gel electrophoresis, a 35-kD protein is absent in the *synag 7* sample and a novel 27-kD protein appears (Pupillo et al., 1988).

The data presented above suggest the model shown in Fig. 1, in which the chemotactic response proceeds from the chemoattractant receptor through $G\alpha 2$ to cause activation of appropriate effectors. The signaling response, however, appears to be dependent on the activation of two GTP-dependent activities, $G\alpha 2$ and a second GTP-dependent activity which regulates the catalytic component of adenylate cyclase. This activity becomes uncoupled from cAMP receptor stimulation when membranes are prepared.

TABLE I
In Vitro Adenylate Cyclase Activity of *Synag* Mutants 5, 9, 7, and 24 Compared with Wild-Type Cells

Additions	WT	Synag 5	Synag 9	Synag 24	Synag 7
Mg^{2+}	1.0	0.8	0.5	0.7	1.0
Mg^{2+} + GTPγS	6.2	1.9	0.8	0.7	2.0
Mn^{2+}	6.1	5.5	3.4	2.8	11.0
WT supernatant + GTPγS	—	—	—	—	16.0

Relative adenylate cyclase activity was measured in the presence of 1 mM $MgSO_4$ (basal), 5 mM $MnSO_4$ (unregulated), or 1 mM $MgSO_4$ plus 40 μM GTPγS (stimulated) for 1 min at 22°C. Adenylate cyclase activity of stimulated *Synag 7* was also measured after preincubation for 10 min at 0°C with a high-speed supernatant of wild-type cells (WT).

This model is reminiscent of the activation of adenylate cyclase in human polymorponuclear leukyocytes (PMNs). In these cells, stimulation with the β-agonist isoproterenol or prostaglandin E_1 (PGE_1) causes sustained elevations of intracellular cAMP, while the chemoattractant *f*-Met-Leu-Phe causes transient cAMP increases. The *f*-Met-Leu-Phe response is dependent on intracellular calcium increases, however, while the β-agonist or PGE_1 response is not. When membranes are prepared from PMNs, the *f*-Met-Leu-Phe–stimulated adenylate cyclase activity is uncoupled. Isoproterenol and PGE_1 can still elicit cAMP elevations. Although uncoupled from adenylate cyclase activation, the chemoattractant receptor still appears to be coupled to a G protein in membranes, which is suggested by the regulation of *f*-Met-Leu-Phe binding by guanine nucleotides (Verghese et al., 1985).

Since addition of inositol 1,4,5-triphosphate to saponin-treated *fgd A* allele HC85 cells restores cAMP stimulation of adenylate cyclase activity, the mobilization of intracellular calcium may be the signal that couples $G\alpha 2$ to adenylate cyclase activation. Stimulation by cAMP is still required for adenylate cyclase activation under these conditions, implying that cAMP stimulation must either proceed through another cAMP receptor, as the chemoattractant receptor is not coupled to its G protein in HC85 cells, or that the receptor can interact with another G protein (Van Haastert, P., personal communication). Clarification of this issue and others presented above awaits further investigation.

Molecular Components of the Signal Transduction Pathways

For several of the components in the model presented in Fig. 1, cDNAs have been isolated and the primary sequence of the proteins has been determined. Homology of these components to those identified in other organisms helps to identify the essential elements in G protein–linked signal transduction and stresses the conservation of these pathways.

The proposed model for the cell-surface cAMP receptor, seven transmembrane domains with the amino terminus on the extracellular side of the membrane (Klein et al., 1988), is topologically similar to the G protein–linked receptors, such as the β-adrenergic receptor (Dohlman et al., 1987) and rhodopsin (Hargrave, 1986). The cytoplasmic carboxy-terminal domain contains several stretches with abundant serine residues, which are the postulated sites of cAMP-induced phosphorylation. As discussed above, this modification has been correlated with adaptation of the cells to the cAMP stimulation.

The primary sequences determined from two G protein α subunit cDNAs, designated Gα1 and Gα2, show 45% identity to each other and to G_i α subunits from rat (Pupillo et al., 1989). Northern blot analysis demonstrates that expression of Gα2, which mediates cAMP stimulation of the chemoattractant receptor, is developmentally regulated and its time course correlates with expression of the cAMP receptor and with the cAMP-stimulated responses described above (Kumagai et al., 1989). The time course of expression of Gα1 does not correlate with guanine nucleotide–regulated adenylate cyclase activity (Kumagai et al., 1989), making it a poor candidate for the G protein playing that role. This suggests that *Dictyostelium* may express other G proteins that have not yet been identified. Based on predictions of a G protein α subunit three-dimensional structure made by Masters et al. (1986), the putative GTP-binding sites in both Gα1 and Gα2 are identical to those in all other G proteins whose amino acid sequence has been determined, except for Gαz (Fong et al., 1988). The *fgd A* allele HC112 contains a point mutation in one of these sites in Gα2 that is analogous to a transforming mutation in the *ras* oncogene product p21 (Pitt, G. S., R. E. Gundersen, and P. N. Devreotes, manuscript in preparation).

The cDNA for one G protein β subunit, which shares 63% identity to the human β-2-subunit, has also been isolated (Lilly, P., and P. N. Devreotes, manuscript in preparation). Northern analysis shows that this protein is constitutively expressed. Whether other β subunits are expressed, whose time course of expression parallels the expression of the identified α subunits, or whether one β subunit can interact with all the α subunits, is being explored.

The recent development of two techniques promises exciting advances in understanding G protein–linked signal tranduction in *Dictyostelium*. De Lozanne and Spudich (1987) described the conditions for the disruption of the myosin heavy chain gene by homologous recombination. The ability to target specific genes for disruption whose products are integral parts of the signal transduction pathway will allow further defining of the molecular components and the biochemistry of these pathways. Dynes and Firtel (1989) have achieved high efficiency transformation of *Dictyostelium* by electroporation, allowing them to cure a thymidine auxotroph by complementation with a library constructed in a plasmid which remains extrachromosomal. This allowed them to rescue the plasmid and identify the gene that complements the mutation. Although the size of inserts in this vector is currently limited to relatively small pieces, future work may allow the identification of the gene product for almost any mutant.

Acknowledgments

This research was supported by Public Health Service grants GM-28007 and GM-34933 to P. N. Devreotes. G. S. Pitt was supported in part by a short-term research training grant (National Institutes of Health, 5T35HL-07606) and by a Medical Scientist Training Program grant (National Insitutes of Health, 5T32GM-07309)

References

Coukell, M. B. 1975. Parasexual genetic analysis of aggregation-deficient mutants of *Dictyostelium discoideum*. *Molecular General Genetics*. 142:119–135.

Darmon, M., J. Barra, and P. Brachet. 1978. The role of phosphodiesterase in aggregation of *Dictyostelium discoideum*. *Developmental Genetics*. 3:283–297.

De Lozanne, A., and J. Spudich. 1987. Disruption of the *Dictyostelium* myosin heavy chain by homologous recombination. *Science*. 236:1086–1091.

Dohlman, H. G., M. G. Caron, and R. J. Lefkowitz. 1987. A family of receptors couple to guanine nucleotide regulatory proteins. *Biochemistry*. 26:2657–2664.

Dynes, J. L., and R. A. Firtel. 1989. Molecular complementation of a genetic marker in *Dictyostelium* using a genomic DNA library. *Proceedings of the National Academy of Sciences*. 86:7966–7970.

Fong, H. K. W., K. K. Yoshimoto, P. Eversole-Cire, and M. I. Simon. 1988. Identification of a GTP-binding protein α-subunit that lacks an apparent ADP-ribosylation site for pertussis toxin. *Proceedings of the National Academy of Sciences*. 85:3066-3070.

Gundersen, R. E., and P. N. Devreotes. 1990. In vivo receptor-mediated phosphorylation of a G protein in *Dictyostelium*. *Science*. In press.

Hargrave, P. A. 1986. Molecular dynamics of the rod cell. *In* The Retina. R. Adler and D. Farber editors. Academic Press, New York. 207–237.

Kesbeke, F., B. E. Snaar-Jagalska, and P. J. M. Van Haastert. 1988. Signal transduction in *Dictyostelium discoideum fgd A* mutants with a defective interaction between surface cAMP receptors and a GTP-binding regulatory protein. *Journal of Cell Biology*. 107:521-528.

Klein, P. S., T. J. Sun, C. L. Saxe III, A. R. Kimmel, R. L. Johnson, and P. N. Devreotes. 1988. A chemoattractant receptor controls development in *Dictyostelium discoideum*. *Science*. 241: 1467–1472.

Kumagai, A., M. Pupillo, R. Gundersen, R. Miake- Lye, P. N. Devreotes, and R. A. Firtel. 1989. Regulation and function of Gα protein subunits in *Dictyostelium*. *Cell*. 57:265–275.

Masters, S. B., R. M. Stroud, and H. R. Bourne. 1986. Family of G protein α chains: amphipathic analysis and predicted structure of functional domains. *Protein Engineering*. 1:47–54.

Pupillo, M., P. Klein, R. Vaughan, G. Pitt, P. Lilly, T. Sun, P. Devreotes, A. Kumagai, and R.

Firtel. 1988. cAMP receptor and G-protein interaction control development in *Dictyostelium*. *Cold Spring Harbor Symposia on Quantitative Biology*. 53:657–665.

Pupillo, M., A. Kumagai, G. S. Pitt, R. A. Firtel, and P. N. Devreotes. 1989. Multiple α subnunits of guanine nucleotides binding proteins in *Dictyostelium, Proceedings of the National Academy of Sciences*. 86:4892–4897.

Theibert, A., and P. N. Devreotes. 1986. Surface receptor-mediated activation of adenylate cyclase in *Dictyostelium. Journal of Biological Chemistry*. 261:15121–15125.

Van Haastert, P. J., R. J. de Wit, P. M. Janssens, F. Kesbeke, and J. DeGoede. 1986. G-protein–mediated interconversions of cell-surface cAMP receptors and their involvement in excitation and desensitization of guanylate cyclase in *Dictyostelium discoideum. Journal of Biological Chemistry*. 261:6904–11.

Vaughan, R. A., and P. N. Devreotes. 1988. Ligand-induced phosphorylation of the cAMP receptor in *Dictyostelium discoideum. Journal of Biological Chemistry*. 263:14538–14543.

Verghese, M. W., K. Fox, L. C. McPhail, and R. Snyderman. 1985. Chemoattractant-elicited alterations of cAMP levels in human polymorphonuclear leukocytes require a Ca^{2+}-dependent mechanism which is independent of transmembrane activation of adenylate cyclase. *Journal of Biological Chemistry*. 260:6769–6775.

Chapter 11

Immunological and Molecular Biological Analysis of the Regulation and Function of Muscarinic Acetylcholine Receptors and G Proteins

Neil M. Nathanson

Department of Pharmacology, SJ-30, University of Washington, Seattle, Washington 98195

Introduction

Muscarinic acetylcholine receptors (mAChR) are thought to be the major type of cholinergic receptor in the central nervous system, where they have been implicated in memory and learning, arousal, control of movement, etc. Muscarinic receptors are also present in peripheral organs and nerves, where they regulate such processes as the rate and force of beating of the heart, smooth muscle contraction, secretion from exocrine glands, and synaptic transmission in parasympathetic and sympathetic ganglia. The physiological actions of mAChR are mediated both by the regulation of the activities of enzymes involved in the production of second messengers and by the direct modulation of ion channel activity. All of the physiological actions of the mAChR are thought to result from their interaction with the G proteins, which can couple mAChR both to such enzymes as adenylate cyclase or phospholipase C and to such ion channels as the atrial inward-rectifying potassium channel (see Nathanson, 1987, for review). This chapter will describe some recent studies from our laboratory on the regulation of expression of muscarinic receptors and G proteins in the heart, and on the properties of receptors expressed from cloned genes in cells that normally do not express the mAChR.

Ontogenesis of the mAChR-mediated Negative Chronotropic Response in the Chick Heart during Embryonic Development

The embryonic chick heart is an attractive system for the study of regulation of expression and function of mAChR during development because of the wealth of anatomical and physiological data available, the relatively low cost, the ease in incubating and using large numbers of embryos, and the relative ease in obtaining cultures of cardiac cells. Very early in embryonic development (3–4-d-old embryos), chick atria are much less responsive to the mAChR-mediated negative chronotropic response than atria isolated from older embryos, even though there are similar numbers of mAChR binding sites in the embryonic chick heart throughout development (Pappano and Skowronek, 1974; Galper et al., 1977; Renaud et al., 1980). Halvorsen and Nathanson (1984) carried out binding studies which demonstrated that the regulation by guanine nucleotides of agonist binding to the mAChR, an effect which is a biochemical manifestation of the interaction of the receptor with its G proteins, was greatly reduced at 4 d of development compared with older ages. They also found that much higher concentrations of guanine nucleotides were required for the regulation of agonist binding by guanine nucleotides and for guanine nucleotide–dependent inhibition of adenylate cyclase activity at 4 d compared with 5 and 8 d of development. These results suggested that the interaction of the mAChR with G proteins was reduced in atria membranes at 4 d of development, perhaps because the levels of G proteins were decreased at early stages of development.

Halvorsen and Nathanson (1984) then directly examined the G protein α subunits by islet-activating protein (IAP)–catalyzed [^{32}P]ADP-ribosylation and autoradiography after one- and two-dimensional gel electrophoresis. Two labeled polypeptides with apparent molecular weights of 41,000–42,000 and 39,000 were observed that were initially assumed to represent two forms of $G_{i\alpha}$. However, subsequent studies demonstrated that after partial proteolysis these two polypeptides yielded nonidentical peptide maps (Martin et al., 1987). Furthermore, immunoblot analysis (Luetje et al., 1987) using antibodies specific for various G protein subunits indicated that the

41,000–42,000 species corresponded to one or more forms of $G_{i\alpha}$, and the 39,000 band contained both $G_{o\alpha}$ and a distinct form of $G_{i\alpha}$. Both molecular weight species could be further resolved using two-dimensional gel electrophoresis into two distinct isoelectric forms (Halvorsen et al., 1984). During embryonic development the acidic form of the 39,000 kD polypeptide increased from 16% of the total IAP-labeled material at day 4 to 55% of the total IAP-labeled material at day 8. There were thus both structural and functional changes in the chick atrial G proteins which occurred at the same time in embryonic development as the ontogenesis of the ability of the mAChR to regulate beat rate, suggesting that these alterations could be responsible for the increased mAChR-mediated negative chronotropic response. Because there was convincing electrophysiological evidence that the mAChR-mediated negative chronotropic response did not involve changes in the level of intracellular cAMP or other diffusible cytosolic second messengers, these results raised the possibility that G proteins might directly couple mAChR to activation of the potassium channel responsible for the negative chronotropic response. Subsequent biochemical, electrophysiological, and molecular biological experiments have shown that G proteins can couple muscarinic and other receptors to the inwardly rectifying potassium channels in the heart, although the identity of the subunits and the precise molecular mechanisms involved remain a topic of controversy (Martin et al., 1985; Pfaffinger et al., 1985; Codina et al., 1987; Logothetis et al., 1988).

Regulation of Muscarinic Receptors and G Proteins in Chick Heart during Embryonic Development

To determine if the levels of signal transduction proteins are regulated in the chick heart in a tissue-specific fashion during embryonic development, Luetje et al. (1987) used a quantitative immunoblot technique to measure the levels of $G_{o\alpha}$, $G_{i\alpha}$, and G_β in atria and ventricles at various stages of development, and also determined the number of mAChR binding sites using a radioligand filter binding assay. At embryonic day 10, the amounts of $G_{o\alpha}$, G_β, and mAChR were similar in atria and ventricles. The levels of these three polypeptides in the atria increased by 46, 80, and 61%, respectively, during the next 5 d of embryonic development. In contrast, the level of $G_{i\alpha}$ was 44% higher in atria than ventricles at day 10, and declined over the next several days. The levels of these polypeptides in the ventricles did not change during this period of development. There were also tissue-specific changes in both guanine nucleotide–mediated and mAChR-mediated inhibition of adenylate cyclase activity during this time. There was a loss of GTP-mediated inhibition of basal adenylate cyclase activity in the atria as well as differences in mAChR-mediated inhibition of adenylate cyclase activity between atria and ventricles. There are thus developmentally regulated tissue-specific alterations in both the expression and function of mAChR and G proteins in the chick heart. This period of development corresponds to the time at which the cholinergic innervation of the chick heart becomes functionally established. Interestingly, Kirby and Aronstam (1983) reported that they could block the atrial-specific increase in mAChR number by blocking the onset of functional cholinergic innervation using the administration of atropine. These results raise the possibility that the expression of both mAChR and the G proteins required for its action may be required in a coordinate tissue-specific fashion by the establishment of cholinergic innervation of the heart.

Regulation of G Protein Expression in Mammalian Heart during Development

Several groups have carried out ADP-ribosylation experiments with rat heart membranes and reported observing only a single labeled polypeptide after one-dimensional sodium dodecylsulfate gel electrophoresis; it was therefore suggested that rat heart contained only $G_{i\alpha}$ and not $G_{o\alpha}$. In addition, the levels of the IAP-labeled substrate varied both in vivo and in vitro in a manner that raised the possibility that the expression of $G_{i\alpha}$ was developmentally regulated in rat heart by innervation (Steinberg et al., 1985; Murakami and Yasuda, 1986). We used antibody and cDNA probes to measure the levels of G protein polypeptides and mRNA in neonatal and adult rat atria and ventricles by quantitative immunoblot, Northern blot, and quantitative RNA dot blot analyses (Luetje et al., 1988). The expression of $G_{o\alpha}$ was regulated in a dramatic cell-specific fashion: there was 5.2-fold more $G_{o\alpha}$ polypeptide in adult atria than in adult ventricles, and 3.4-fold more $G_{o\alpha}$ mRNA in adult atria than in adult ventricles. There was also 2.8-fold more G_β polypeptide in adult atria than ventricles, although the two tissues contained similar amounts of G_β mRNA. During development from birth to adulthood, the ventricular levels of $G_{i\alpha}$ polypeptide and mRNA, G_β mRNA, and $G_{o\alpha}$ mRNA, and the atrial levels of G_β and $G_{i\alpha}$-2 mRNA, all decreased. The level of $G_{i\alpha}$-3 mRNA in the atria increased during development. Thus, there are tissue-specific and developmentally regulated differences in the expression of G protein subunits in rat heart. Because the innervation of the rat heart occurs postnatally, it is possible that some or all of these developmentally regulated changes result from innervation. Experiments to test the role of innervation in the regulation of G protein expression in both chick and rat hearts are currently in progress.

Cloning and Expression of Mammalian Muscarinic Receptor Genes

At least five subtypes of mAChR have been shown to exist in mammals and to be the products of separate genes (Bonner, 1989). We used a cDNA clone encoding the porcine m2 mAChR to isolate a genomic clone encoding the mouse m1 receptor (Shapiro et al., 1988). The gene encoded a polypeptide of 460 amino acids whose deduced sequence was very similar to m1 mAChRs from other species; for example, there were 13 (mostly conservative) amino acid changes from the procine m1 receptor. The mouse m1 gene, like all other vertebrate mAChR genes described to date, did not contain introns in the coding region of the gene. To test if this putative mAChR gene encoded a functional receptor, the coding region of the gene was inserted into a mammalian expression vector under the transcriptional control of the mouse metallothionein promoter, and was expressed in stably transfected mouse Y1 adrenal and L fibroblast cells. Control experiments demonstrated that untransfected Y1 or L cells did not express detectable levels of mAChR binding sites and that incubation with muscarinic agonists did not alter phosphoinositide hydrolysis or intracellular cAMP levels. The mouse m1 receptor when expressed in Y1 cells exhibited the high affinity binding for the antagonists quinuclidinyl benzilate, atropine, and pirenzepine expected for a m1 receptor. The m1 receptor was physiologically active, as evidenced by its ability to activate phospholipase C. In Y1 cells expressing the mouse m1 receptor, carbachol stimulated phosphoinositide hydrolysis a maximum of 20–40-fold. The stimulation in phosphoinositide hydrolysis evoked by 1 mM carbachol could be blocked

90% by coincubation with 1 μM atropine, confirming the muscarinic specificity of this response. Carbachol did not cause an inhibition of cAMP accumulation in either type of cell; in fact, cAMP levels in Y1 cells increased in response to carbachol. To ensure that this apparent selectivity of functional coupling was due to the m1 receptor and not the Y1 cells, two types of control experiments were performed. First, the functional specificity of the m1 receptor was determined after its expression in a second cell line, mouse L cells. Incubation of transfected L cells with carbachol caused a 2.5-fold increase in phosphoinositide hydrolysis that could be completely blocked by atropine. There was no change in intracellular cAMP levels after incubation of the cells with atropine. Thus, the m1 receptor when expressed in either Y1 or L cells was able to activate phospholipase C but could not inhibit adenylate cyclase.

As an additional control, a cDNA clone encoding the porcine m2 receptor was also expressed in Y1 cells. The m2 receptor caused inhibition of adenylate cyclase but did not activate phospholipase C. Incubation with carbachol caused a 30–65% inhibition in forskolin-stimulated cAMP accumulation but did not cause a significant change in phosphoinositide hydrolysis. Thus, the m1 and m2 receptors mediate different physiological responses when expressed in Y1 cells. Furthermore, the different functional responses mediated by the two receptor subtypes exhibit different sensitivities to ADP-ribosylation by IAP. Preincubation of Y1 cells with IAP completely eliminated m2-mediated inhibition of forskolin-stimulated cAMP accumulation, but did not block m1-mediated stimulation of phosphoinositide hydrolysis. These results demonstrate that the m1 and m2 receptors mediate different functional responses in Y1 adrenal cells by interacting with different G proteins.

Deletion Analysis of Function and Regulation of the m1 Receptor

The third putative cytoplasmic loop in the G protein–coupled receptors is thought to play a key role in the interaction of these receptors with G proteins to produce physiological responses. Thus, small deletions in this portion in the β-adrenergic receptor causes a loss of coupling of the receptor to G_S, and switching this region between the α- and β-adrenergic receptors switches the physiological response of the chimeric receptor (Kobilka et al., 1988). The third cytoplasmic loop also appears to contain the amino acid residues that determine the specificity of mAChR coupling; switching the entire third cytoplasmic loop between the m1 mAChR and the m2 mAChR results in a switch in the type of ion channel the hybrid receptors could regulate after expression in oocytes (Kubo et al., 1988).

We have constructed deletion mutations in the third cytoplasmic loop of the mouse m1 receptor gene to more precisely localize which portion of the third cytoplasmic loop determines the functional coupling of the m1 mAChR (Shapiro and Nathanson, 1989). These mutant mAChRs were expressed in Y1 cells and the ability of the mutant receptors to stimulate phospholipase C activity was determined. Two mutants were made in which either 64 or 123 of the 156 amino acids in the third cytoplasmic loop were deleted. The mutant receptors when expressed in Y1 cells had affinities for agonists and antagonists that were identical to that of the wild-type receptor. In addition, the mutant and wild-type receptors were indistinguishable in their ability to stimulate phosphoinositide hydrolysis; both the maximal extent of stimulation and the concentrations of agonist required to evoke the response were identical in the wild-type and mutant receptors. These results combined with the results of Kubo et al. (1988)

demonstrate that the membrane proximal portions of the third cytoplasmic loop determine the functional specificity of muscarinic receptor action.

While deletion of 75% of the third cytoplasmic loop did not alter coupling of the m1 receptor to phospholipase C, it had an interesting effect on the regulation of receptor expression after long-term agonist exposure. The mutant receptor missing 123 amino acids in the loop exhibited a diminished susceptibility to agonist-induced downregulation without altering short-term agonist-induced receptor internalization. These results suggest that this region of the receptor may be involved in targeting internalized receptors to lyosomes for degradation or that it may represent a target site for intracellular proteases.

Cloning and Functional Expression of a *Drosophila* Muscarinic Receptor

Acetylcholine is a major neutrotransmitter in the *Drosophila* nervous system, and mAChR have been identified in *Drosophila* using radioligand binding (Haim et al., 1979). We have isolated cDNA and genomic clones encoding a mAChR from *Drosophila*, which we have called Dm1 (Shapiro et al., 1989). The cDNA clone encodes a polypeptide of 722 amino acids, which is much larger than any of the previously cloned mammalian mAChR (which range from 460 to 590 amino acids). While the mammalian mAChR genes do not contain introns in their coding regions, the *Drosophila Dm1* gene contains three introns in the coding region, all in the portion of the gene encoding the third putative cytoplasmic loop. The deduced amino acid sequence (excluding the highly variable third cytoplasmic loop is 45–50% identical to the mammalian m1, m2, m3, m4, and m5 receptors. The *Drosophila* mAChR is much more similar to the mammalian m1, m3, and m5 receptors than to the m2 and m4 receptors in the membrane proximal portions of the third cytoplasmic loop, suggesting that it would preferentially couple to stimulation of phospholipase C activity rather than inhibition of adenylate cyclase activity. The presence of introns in the coding region of the gene raises the possibility that multiple types of mAChR may arise in *Drosophila* not only from additional (and as yet unknown) genes but also by alternative splicing of the *Dm1* gene.

The cDNA clone encoding the *Drosophila Dm1* receptor was expressed in Y1 mouse adrenal cells to ensure that it encoded a functional mAChR and to compare its properties with those of the mammalian mAChR previously expressed in this cell line. The *Dm1* protein exhibited high affinities for the muscarinic antagonists quinuclidinyl benzilate and atropine, which were similar to those reported by Haim et al. (1979) for the binding of muscarinic ligands to receptors in homogenates prepared from *Drosophila* heads, thus confirming that *Dm1* encoded a muscarinic receptor.

As noted above, the high degree of amino acid identity of the *Dm1* receptor to the mammalian m1, m3, and m5 receptors in the membrane proximal portions of the third cytoplasmic loop suggested that the *Drosophila* mAchR should preferentially couple to phospholipase C rather than to adenylate cyclase. As was previously found with Y1 cells expressing the mammalian m1 receptor, addition of carbachol to Y1 cells expressing *Dm1* increased phosphoinositide hydrolysis 18–20-fold over control levels and increased forskolin-stimulated cAMP accumulation. Thus, the *Drosophila Dm1* mAChR is both structurally and functionally homologous to the mammalian m1 mAChR.

The isolation of the mAChR gene from *Drosophila* should allow a genetic analysis of the role of the receptor in the whole animal. The introduction and expression of wild-type and mutant mAChR genes into *Drosophila* deleted of the *Dm1* locus will be combined with behavioral and biochemical studies to determine the physiological significance and functional role of the mAChR in *Drosophila*.

Cloning and Expression of an Avian Muscarinic Receptor

There are many similarities between mAChR in mammalian and chick hearts. For example, mAChR are coupled to adenylate cyclase, potassium channels, and phospholipase C in both avian and mammalian hearts. However, there are also a number of important differences between avian and mammalian cardiac mAChR. The two types of receptor can be distinguished pharmacologically, with the mammalian cardiac mAChR exhibiting a much lower affinity for the M1-specific antagonist pirenzepine than chick heart mAChR (Brown et al., 1984; Kwatra et al., 1989). The receptors are also immunologically distinct: we have isolated a series of monoclonal antibodies that recognize mAChR from pig, rat, and mouse hearts but which do not recognize the chick heart mAChR (Subers et al., 1988). Finally, peptide mapping studies of mAChR purified from mammalian and chick hearts demonstrate that the receptor proteins are different (Kwatra et al., 1989). We have isolated and sequenced the gene encoding the chick heart mAChR to determine the molecular basis for the differences between mammalian and chick cardiac mAChR and to have nucleic acid probes to use to study the regulation of mAChR gene expression in the chick heart (Tietje et al., 1990).

A chicken genomic library was screened at moderate stringency with a hybridization mix consisting of full-length nick-translated mouse genomic m1 and porcine m2 cDNA clones. A clone that hybridized strongly to both probes was sequenced and shown to encode a polypeptide of 490 amino acids with a high degree of homology to the mammalian mAChR. Excluding the highly variable third putative cytoplasmic loop, this putative chick mAChR exhibited 52, 70, 54, 83, and 53% amino acid identity to the mammalian m1, m2, m3, m4, and m5 receptors, respectively. We therefore refer to this chick mAChR as cm4, for chick m4 mAChR. Northern blot analysis demonstrated that the cm4 gene hybridizes to a 2.8 kb RNA expressed in chick brain and heart. When expressed in Chinese hamster ovary (CHO) cells or mouse Y1 adrenal cells, this mAChR exhibited affinities for quinuclidinyl benzilate, atropine, pirenzepine, and AF-DX 116 that are similar to the values previously reported for the mAChR in chick heart (Brown et al., 1984; Kwatra et al., 1989). These results demonstrate that the cm4 gene encodes a mAChR normally expressed in chick heart, which when expressed in heterologous cell types exhibits the ligand binding properties normally seen in situ in the chick heart.

The functional specificity of the cm4 receptor when expressed in Y1 and CHO cells was also examined. When expressed in Y1 cells, the cm4 receptor exhibited the same specificity for functional activity as previously observed for the mammalian cardiac (m2) receptor; there was agonist-mediated inhibition of adenylate cyclase activity but not agonist-mediated stimulation of phosphoinositide hydrolysis. In contrast, the cm4 receptor, like the mammalian m2 and m4 receptors, coupled to both inhibition of adenylate cyclase and stimulation of phosphoinositide hydrolysis when expressed in CHO cells. Thus, the chick heart mAChR has a high degree of both sequence and functional homology to the mammalian cardiac mAChR. In addition, these results demonstrate that the functional specificity of an individual mAChR subtype can be determined not only by the amino acid sequence encoded by its gene but also by the type of cell that gene is expressed in.

Acknowledgments

The work described here was supported by a Grant-in-Aid from the American Heart Association, by National Institutes of Health grants GM-07270, HL-30639, HL-07312, and NS-26920, by the U.S. Army Research Office, and by grant IN-26-29 from the American Cancer Society. N. M. Nathanson is an Established Investigator of the American Heart Association.

References

Bonner, T. I. 1989. the molecular basis of muscarinic receptor diversity. *Trends in Neurosciences.* 12:148–151.

Brown, J. H., D. Goldstein, and S. B. Masters. 1984. The putative M1 muscarinic receptor does not regulate phosphoinositide hydrolysis. Studies with pirenzepine and McN-A343 in chick heart and astrocytoma cells. *Molecular Pharmacology.* 27:525–531.

Codina, J., A. Yatani, D. Grenet, A. M. Brown, and L. Birnbaumer. 1987. The α subunit of the GTP binding protein G_k opens atrial potassium channels. *Science.* 236:442–445.

Galper, J. B., W. Klein, and W. A. Catterall. 1977. Muscarinic acetylcholine receptors in developing chick heart. *Journal of Biological Chemistry.* 252:8692–8699.

Haim, N., S. Nahum, and Y. Dudai. 1979. Properties of a putative muscarinic cholinergic receptor from *Drosophila melanogaster. Journal of Neurochemistry.* 32:543–552.

Halvorsen, S. W., and N. M. Nathanson. 1984. Ontogenesis of physiological responsiveness and guanyl nucleotide sensitivity of cardiac muscarinic acetylcholine receptors during chick embryonic development. *Biochemistry.* 23:5813–5821.

Kirby, M. L., and R. S. Aronstam. 1983. Atropine-induced alterations of normal development of muscarinic receptors in the embryonic chick heart. *Journal of Molecular and Cellular Cardiology.* 15:685–696.

Kobilka, B. K., T. S. Kobilka, K. Daniel, J. W. Regan, M. G. Caron, and R. J. Lefkowitz. 1988. Chimeric $\alpha 2$-, $\beta 2$-adrenergic receptors: delineation of domains involved in effector coupling and ligand specificity. *Science.* 240:1310–1316.

Kubo, T., H. Bujo, I. Akiba, J. Nakai, M. Mishina, and S. Numa. 1988. Localization of a region of the muscarinic acetylcholine receptor involved in selective effector coupling. *Nature.* 241:119–125.

Kwatra, M. M., J. Ptasienski, and M. M. Hosey. 1989. The procine heart M2 receptor: agonist-induced phosphorylation and comparison of properties with the chick heart receptor. *Molecular Pharmacology.* 35:553–558.

Logothetis, D. E., D. Kim, J. K. Northrup, E. J. Neer, and D. E. Clapham. 1988. Specificity of action of guanine nucleotide-binding regulatory protein subunits on the cardiac muscarinic K^+ channel. *Proceedings of the National Academy of Sciences.* 85:5814–5818.

Luetje, C. W., P. Gierschik, G. Milligan, C. Unson, A. Spiegel, and N. M. Nathanson. 1987. *Biochemistry.* 26:4876–4884.

Luetje, C. W., K. M. Tietje, J. L. Christian, and N. M. Nathanson. 1988. Differential tissue expression and developmental regulation of guanine nucleotide binding regulatory proteins and their mRNAs in rat heart. *Journal of Biological Chemistry.* 263:13357–13365.

Martin, J. M., D. D. Hunter, and N. M. Nathanson. 1985. Islet activating protein inhibits physiological responses evoked by cardiac muscarinic acetylcholine receptors. Role of GTP-binding proteins in regulation of potassium permeability. *Biochemistry.* 23:5813–5821.

Martin, J. M., E. M. Subers, S. W. Halvorsen, and N. M. Nathanson. 1987. Functional and physical properties of atrial and ventricular GTP-binding proteins: relationship to muscarinic acetylcholine receptor mediated responses. *Journal of Pharmacology and Experimental Therapeutics.* 240:683–688.

Murakami, T., and H. Yasuda. 1986. Rat heart cell membranes contain three substrates for cholera-toxin–catalyzed ADP-ribosylation and a single substrate for pertussis toxin–catalyzed ADP-ribosylation. *Biochemical and Biophysical Research Communications.* 138:1355–1361.

Nathanson, N. M. 1987. Molecular properties of the muscarinic acetylcholine receptor. *Annual Review of Neuroscience.* 10:195–236.

Pappano, A. J., and A. C. Skowronek. 1974. Reactivity of chick embryo heart to cholinergic agonists during ontogenesis: decline in desensitization at the onset of cholinergic transmission. *Journal of Pharmacology and Experimental Therapeutics.* 191:109–118.

Pfaffinger, P. J., J. M. Martin, D. D. Hunter, N. M. Nathanson, and B. Hille. 1985. GTP-binding proteins couple cardiac muscarinic receptors to a K channel. *Nature.* 317:536–538.

Renaud, J. F., J. Barhanin, D. Cavey, M. Fosset, and M. Lazdunski. 1980. Comparative properties of the *in ovo* and *in vitro* differentiation of the muscarinic cholinergic receptor in embryonic heart cells. *Developmental Biology.* 78:184–200.

Shapiro, R. A., and N. M. Nathanson. 1989. Deletion analysis of the m1 muscarinic acetylcholine receptor: effects on phosphoinositide metabolism and down-regulation. *Biochemistry.* 28:8946–8950.

Shaprio, R. A., N. M. Scherer, B. A. Habecker, E. M. Subers, and N. M. Nathanson. 1988. Isolation, sequence, and functional analysis of the M1 muscarinic acetylcholine receptor. *Journal of Biological Chemistry.* 263:18397–18403.

Shaprio, R. A., B. T. Wakimoto, E. M. Subers, and N. M. Nathanson. 1989. Characterization and functional expression in mammalian cells of genomic and cDNA clones encoding a *Drosophila* muscarinic acetylcholine receptor. *Proceedings of the National Academy of Sciences.* 86:9039–9043.

Steinberg, S. F., E. D. Drugge, J. P. Bilzezikian, and R. B. Robinson. 1985. Acquisition by innervated myocytes of a pertussis toxin–specific regulatory protein linked to the α_1-receptor. *Science.* 230:186–188.

Subers, E. M., W. C., Liles, C. W. Luetje, and N. M. Nathanson. 1988. Biochemical and immunological studies on the regulation of cardiac and neuronal muscarinic acetylcholine receptor number and function. *Trends in Pharmacological Sciences Supplement: Subtypes of Muscarinic Receptors. III* 25–28.

Tietje, K. M., P. S. Goldman, and N. M. Nathanson. 1990. Cloning and functional analysis of a gene encoding a novel muscarinic acetylcholine receptor expressed in chick heart and brain. *Journal of Biological Chemistry.* 265:2828–2834.

Chapter 12

**Structural and Functional Studies
of the G_o Protein**

Eva J. Neer

*Department of Medicine, Brigham and Women's Hospital and
Harvard Medical School, Boston, Massachusetts 02115*

Introduction

The guanine nucleotide–binding proteins, which couple a variety of hormones, neurotransmitters, and chemotactic and sensory signals to cellular enzymes and ion channels, are made up of three polypeptides, α, β, and γ. The α chain binds and hydrolyzes guanine nucleotides. Most α subunits can be substrates for modification by a bacterial toxin, either cholera toxin or pertussis toxin (for recent reviews, see Gilman, 1987; Lochrie and Simon, 1988; Neer and Clapham, 1988). The set modified by pertussis toxin includes the two transducins (α_t; retinal proteins that couple rhodopsin to a cGMP phosphodiesterase), the three α_i proteins (α_{i-1}, α_{i-2}, α_{i-3}) and a closely related protein, α_o (Neer et al., 1984; Sternweis and Robishaw, 1984). The three α_i isoforms are more closely related to each other than to α_t or α_o. Fig. 1 diagrams the degree of similarity between α_i and α_o subunits. The figure compares the sequence of human α_{i-3} to bovine α_o. These two proteins, as well as α_{i-1} and α_{i-2}, are identical in the regions A, C, E, and G that make up the guanine nucleotide–binding site. They are most different at the carboxyl and amino termini and at a region surrounding amino acid 100. There is good evidence that the amino terminal 2-kD region is involved in interactions with $\beta\gamma$

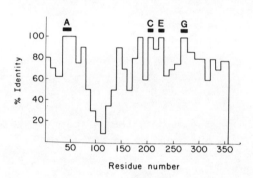

Figure 1. Comparison of amino acid sequences of α_{i-3} and α_o. The number of identical amino acid residues in a given segment was expressed as a percentage and was plotted as a function of the amino acid residue number. The length of segment (11 ± 3 residues) was determined to maximize the identity. The sequence of α subunits can be found in the following references: α_{i-3} (Kim et al., 1988), α_o (Jones and Reed, 1987; Van Meurs et al., 1987).

(Navon and Fung, 1987; Neer et al., 1988b), and that the carboxyl terminal is involved in interactions of the α subunits with receptors (Van Dop et al., 1984; West et al., 1985). The function of the large region around amino acid 100 is not yet known.

Functional Properties of α_o

Although α_o requires detergent for solubilization, once released from the membrane it behaves as a hydrophilic molecule that remains unaggregated in solution without detergent (Huff et al., 1985; Sternweis, 1986; Neer, E. J., unpublished observations). Despite this apparent hydrophilicity, Buss et al. (1987) have shown that α_o is myristoylated at the amino terminus as is α_i, but not α_s, the stimulatory G protein α subunit. Other posttranslational modifications may also occur. Goldsmith et al. (1988) identified a protein immunologically indistinguishable from α_o that has a more basic isoelectric point. The authors suggest this protein may represent a modified α_o.

Although the primary structure of the group of α_i proteins is very similar to that of α_o, the proteins are functionally subtly different. α_o has a more active GTPase activity than α_{i-1} (Neer et al., 1984). It also has a lower affinity for $\beta\gamma$ than does α_{i-1} (Huff and

Neer, 1986). Since the true substrate for ADP-ribosylation by pertussis toxin is not the isolated α subunit, but the $\alpha\beta\gamma$ heterotrimer, a decreased affinity for $\beta\gamma$ is reflected in a lower susceptibility to modification by the toxin. It is not known whether differences in relative affinity of α subunits for $\beta\gamma$ are important determinants of ADP-ribosylation intact cells, although, in principle, they should affect the relative abilities of α subunits to be modified. The functional consequences of ADP-ribosylation are also subtly different for the two α subunits. ADP-ribosylation produces a different conformation in α_o than in α_i, as revealed by differences in patterns of proteolytic cleavage (Winslow et al., 1986).

Many laboratories have tried to show functional differences between α_o and α_i molecules by reconstitution experiments in vitro. One experimental scheme has been to reconstitute hormone receptors with guanine nucleotide–binding protein and to measure receptor/G protein interactions by assessing the effect of guanine nucleotides in ligand binding. Another design is to reconstitute G proteins into membranes and to ask which G protein can couple a particular hormone receptor to a particular effector. The results of such experiments have been variable. In most cases, there is no clear difference between the ability of α_i and α_o to reconstitute a response (Asano et al., 1985; Florio and Sternweis, 1985; Kikuchi et al., 1986; Kurose et al., 1986; Tsai et al., 1987; Logothetis et al., 1988; Ueda et al., 1988; Yatani et al., 1988; Florio and Sternweis, 1989; Haga et al., 1989). However, there are exceptions (for example, Ewald et al., 1989; McFadzean et al., 1989; Ueda et al., 1989). Reconstitution experiments show what interactions might take place, but they do not reveal what the actual function of α_o is in the cell. Although there are a number of clues, this is still a mystery. One idea is that α_o acts to regulate phospholipase C in those cases where phospholipase C activation is pertussis toxin–sensitive (Worley et al., 1986). Another notion is that α_o may regulate ion channels, especially in the central nervous system (Hescheler et al., 1987; Ewald et al., 1988; Van Dongen et al., 1988; Holz et al., 1989). Of course, these possibilities are not mutually exclusive.

Distribution of α_o

The cellular distribution of the group of pertussis toxin substrates varies with its function. Transducin, which transmit visual signals, is located entirely in the retina with transducin 1 found only in rods and transducin 2 found only in cones (Grunwald et al., 1986; Lerea et al., 1986). The α_i family is, for the most part, ubiquitously distributed. In a survey of human cells expressing mRNA for α_{i-2} and α_{i-3}, we found that all cells express some level of these two proteins, although the ratios appear to vary. The most uneven distribution in the α_i family is α_{i-1}, which is apparently lacking from some cell types, including cells of myelocytic origin and some other cells (GH_4 cells, for example). Although the proteins are ubiquitously distributed, each cell type appears to have a characteristic distribution of mRNAs (Kim et al., 1988). The heterogeneity in mRNA is reflective in two-dimensional gel analysis of ADP-ribosylated protein, which revealed a characteristic fingerprint for each cell type (Neer et al., 1988a). The functional consequences of differences in relative abundance of α_i subtypes is not known.

Outside of the retinal cells, the α subunit with the most uneven distribution is α_o, which makes up ~0.5% of the total particulate protein in the brain. It is present in some peripheral tissues at approximately one-tenth the concentration found in brain, and is entirely absent from other peripheral tissues (Huff et al., 1985; Gierschik et al., 1986; Homburger et al., 1987; Asano et al., 1988). Within the brain, the distribution is not uniform, but the protein is highly localized (Worley et al., 1986). If the concentration of α_o is 0.5% of total particulate brain protein, and α_o is, in fact, found to be highly localized in certain brain regions, then in those regions the concentrations must approach 5% of the cell's particulate protein. The amount of $\beta\gamma$ in the central nervous system seems to be sufficient to match all of the α. Therefore, the signaling proteins make up ~10% of the membrane of some cells.

In some cases, α_o appears to be located in defined domains of the cell. In the central nervous system, α_o is located in the neuropil rather than in cell bodies (Worley et al., 1986). Peraldi et al. (1989) found α_o immunoreactivity both in the choroid plexus and in cultured ependymal cells. In these polarized cells, the α_o immunoreactivity was concentrated in the apical plasma membrane. The authors suggest that in these cells, G_o may be involved in signaling pathways related to traffic to or from the cerebrospinal fluid. In chromaffin cells, α_o may be associated with granule membranes as well as the plasma membrane (Toutant et al., 1987).

Regulation of α_o Levels

The abundance of the G_o protein in the brain stands in sharp contrast to the very low level of α_o mRNA. In the family of G proteins, there seems to be a roughly inverse relationship between abundance of mRNA and abundance of protein, so that mRNA for α_s subunits is readily detectable, although the amount of the proteins is very low; the reverse is true for α_o (Brann et al., 1987; Jones and Reed, 1987). The message for α_o is heterogenous with multiple forms having been identified in the brain and peripheral tissues (Price et al., 1989). The functional significance of the heterogeneity in mRNA is not known.

A broad discrepancy between mRNA and protein levels for α_o suggest that α_o may be a particularly stable protein. To investigate this possibility, we studied the degradation rate of α_o in two cell types: GH_4 cells, a clonal pituitary cell line, and cultured neonatal rat cardiocytes. The cells were metabolically labeled with [^{35}S]methionine and the label was then removed and chased with excess cold methionine. α_o was immunoprecipitated using a rabbit polyclonal antibody against brain α_o at various times after the chase. We found that the half-time of degradation of α_o was widely different in the two cell types, being 28 ± 7 h in GH_4 and >72 h in the cardiocytes. Appropriate controls were done to show that all of the α_o was, in fact, being immunoprecipitated in the growing cells and that, in the case of the nondividing cardiocytes, the chase was adequate and other proteins showed more rapid degradation rates than α_o. We measured the mRNA levels for α_o in the two cell types using Northern analysis and found that the mRNA was approximately equal. The steady-state level of the protein in the two cell types was also approximately equal. Since the degradation rates are different, we conclude that the rate of synthesis of α_o in GH_4 cells must be different from cardiocytes. These different rates of synthesis are not proportional to mRNA levels and we suggest, therefore, that translational controls may play a role in determining the steady-state level of the α_o protein (Silbert et al., 1990).

The levels of α_o protein change during development and differentiation. In the chick, heart α_o (or a similar protein) increases during embryonic development of the heart as the heart acquires responsiveness to muscarinic cholinergic agonists (Harlvorsen and Nathanson, 1984; Liang et al., 1986). Changes in amount of the protein also occur during development in rat brain and heart (Asano et al., 1988). Cellular differentiation leads to increases in α_o immunoreactivity in neuroblastoma x glioma hybrid cells (Mullaney and Milligan, 1989). However, the absolute level is always lower than the amount in normal brain (Neer, E. J., unpublished observations).

Conservation of α_o from Mammals to Flies

Like other G protein α subunits, the sequence of α_o is virtually entirely conserved among mammals (Itoh et al., 1986; Jones and Reed, 1987; Van Meurs et al., 1987). The sequence of α_o from *Xenopus laevis* is 89% identical to rat α_o (Olate et al., 1989). Immunologic studies suggested that a similar protein was also present in invertebrates (Homburger et al., 1987).

Despite the hints to its function, the real biologic role of G_o remains elusive. Therefore, it would be extremely useful to study α_o in an organism that can be manipulated genetically. Yeasts have already been extremely productive as model systems, but they have a drawback in that they lack a nervous system. Therefore, we turned to *Drosophila* and asked whether this organism has an analogue of α_o and, if so, whether its cellular distribution is similar to that found in mammals. Using a bovine α_{i-1} cDNA fragment as a probe, Schmidt et al. (1989) have been able to isolate first a partial genomic, and then a cDNA clone that encoded an α subunit that was 81% identical to mammalian α_o protein. (Since presentation of this manuscript, three independent reports of cloning α_o from *Drosophila* have appeared: de Souse et al., 1989; Thambi et al., 1989; Yoon et al., 1989). The α subunit predicted by this sequence was more similar to mammalian α_o than it was to any other mammalian G protein α subunit or to other cloned *Drosophila* subunits. In situ hybridization experiments using a fragment of cDNA from the 3'-untranslated region showed that the *Drosophila* α_o mRNA was located in the cell bodies of the central nervous system, in the thoracic and abdominal ganglia, and in the ovary. In order to determine where the protein was located, we expressed the *Drosophila* α_o in *E. coli* as a fusion protein and used this protein to generate polyclonal antibodies. Immunocytochemical analysis showed that, as in the mammalian nervous system, the *Drosophila* α_o is located primarily in the neuropil of the central nervous system and of the thoracic and abdominal ganglia. The matching sequence and cellular localization suggests that the function of α_o will also be conserved between flies and mammals, and that this simple organism is a suitable model to use in its elucidation (Schmidt et al., 1989).

Five years after the discovery of G_o, there are some clues to its cellular function, but there is neither conclusive evidence identifying its biologic role nor a good explanation for its abundance in the nervous system. The development of immunological and molecular genetic tools for analysis of G_o function promises to allow definition of its role in the nervous system and in other cells.

Acknowledgments

I would like to thank Mrs. Paula McColgan for expertly typing the manuscript.
This work was supported by National Institutes of Health grants GM-36259 and GM-35416.

References

Asano, T., K. Nobuko, R. Semba, and K. Kato. 1988. Ontogeny of the GTP-binding protein G_o in rat brain and heart. *Journal of Neurochemistry.* 51:1711–1716.

Asano, T., M. Ui, and N. Ogasawara. 1985. Prevention of the agonist binding to γ-aminobutyric acid B receptors by guanine nucleotides and islet-activating protein, pertussis toxin, in bovine cerebral cortex. *Journal of Biological Chemistry.* 260:12653–12658.

Brann, M. R., R. M. Collins, and A. Spiegel. 1987. Localization of mRNAs encoding the α-subunits of signal-transducing G-proteins within rat brain and among peripheral tissues. *FEBS Letters.* 222:191–198.

Buss, J. E., S. M. Mumby, P. J. Casey, A. G. Gilman, and B. M. Sefton. 1987. Myristoylated α subunits of guanine nucleotide-binding regulatory proteins. *Proceedings of the National Academy of Sciences.* 84:7493–7497.

de Sousa, S. M., L. L. Hoveland, S. Yarfitz, and J. B. Hurley. 1989. The *Drosophila* G_oα-like G protein gene produces multiple transcripts and is expressed in the nervous system and in ovaries. *Journal of Biological Chemistry.* 264:18544–18551.

Ewald, D. A., I.-H. Pang, P. C. Sternweis, and R. J. Miller. 1989. Differential G protein–mediated coupling of neurotransmitter receptors to Ca^{2+} channels in rat dorsal root ganglion neurons in vitro. *Neuron.* 2:1185–1193.

Ewald, D. A., P. C. Sternweis, and R. J. Miller. 1988. Guanine nucleotide–binding protein G_o induced coupling of neuropeptide Y receptors to Ca^{2+} channels in sensory neurons. *Proceedings of the National Academy of Sciences.* 85:3633–3637.

Florio, V. A., and P. C. Sternweis. 1985. Reconstruction of resolved muscarinic cholinergic receptors with purified GTP-binding proteins. *Journal of Biological Chemistry.* 260:3477–3483.

Florio, V. A., and P. C. Sternweis. 1989. Mechanisms of muscarinic receptor action on G_o in reconstituted phospholipid vesicles. *Journal of Biological Chemistry,* 264:3909–3915.

Gierschik, P., G. Milligan, M. Pines, P. Goldsmith, J. Codina, W. Klee, and A. Spiegel. 1986. Use of specific antibodies to quantitate the guanine nucleotide-binding protein G_o in brain. *Proceedings of the National Academy of Sciences.* 83:2258–2262.

Gilman, A. G. 1987. G-proteins: transducers of receptor-generated signals. *Annual Review of Biochemistry.* 56:615–650

Goldsmith, P., P. S. Backlund, Jr., K. Rossiter, A. Carter, G. Milligan, C. G. Unson, and A. Spiegel. 1988. Purification of heterotrimeric GTP-binding proteins from brain: Identification of a novel form of G_o. *Biochemistry.* 27:7085–7090.

Grunwald, G. B., P. Gierschik, M. Nirenberg, and A. Spiegel. 1986. Detection of α-transducin in retinal rods but not cones. *Science.* 231:856–859.

Haga, K., H. Uchiyama, T. Haga, A. Ichiyama, K. Kangawa, and H. Matsuo. 1989. Cerebral muscarinic acetylcholine receptors interact with three kinds of GTP-binding proteins in a reconstitution system of purified components. *Molecular Pharmacology.* 35:286–294.

Halvorsen, S. W., and N. M. Nathanson. 1984. Ontogenesis of physiological responsiveness and guanine nucleotide sensitivity of cardiac muscarinic receptors during chick embryonic development. *Biochemistry.* 23:5813–5821.

Hescheler, J., W. Rosenthal, W. Trautwein, and G. Schultz. 1987. The GTP-binding protein G_o regulates neuronal calcium channels. *Nature.* 325:445–447.

Holz, G. G., R. M. Kream, A. Spiegel, and K. Dunlap. 1989. G proteins couple of α-adrenergic and GABAb receptors to inhibition of peptide secretion from peripheral sensory neurons. *Journal of Neuroscience.* 9:657–666.

Homburger, V., P. Brabet, Y. Audigier, C. Pantaloni, J. Bockaert, and B. Rouot. 1987. Immunological localization of the GTP-binding protein G_o in different tissues of vertebrates and invertebrates. *Molecular Pharmacology.* 31:313–319.

Huff, R. M., J. M. Axton, and E. J. Neer. 1985. Physical and immunological characterization of a guanine nucleotide-binding protein purified from bovine cerebral cortex. *Journal of Biological Chemistry.* 260:10864–10871.

Huff, R. M., and E. J. Neer. 1986. Subunit interactions of native and ADP-ribosylated α_{39} and α_{41}, two guanine nucleotide binding proteins from bovine cerebral cortex. *Journal of Biological Chemistry.* 261:1105–1110.

Itoh, H., T. Kozasa, S. Nagata, S. Nakamura, T. Katada, M. Ui, S. Iwai, E. Ohtsuka, H. Kawasaki, K. Suzuki, and Y. Kaziro. 1986. Molecular cloning and sequence determination of cDNAs for α subunits of the guanine nucleotide-binding proteins G_s, G_i and G_o from rat brain. *Proceedings of the National Academy of Sciences.* 83:3776–3780.

Jones, D. T., and R. R. Reed. 1987. Molecular cloning of five GTP-binding protein cDNA species from rat olfactory neuroepithelium. *Journal of Biological Chemistry.* 262:14241–14249.

Kikuchi, A., O. Kozawa, K. Kaibuchi, T. Katada, M. Ui, and Y. Takai. 1986. Direct evidence for involvement of a guanine nucleotide-binding protein in chemotactic peptide-stimulated formation of inositol bisphosphate and trisphosphate in differentiated human leukemic (HL-60) cells. *Journal of Biological Chemistry.* 261:11558–11562.

Kim, S. Y., S.-L. Ang, D. Bloch, K. Bloch, Y. Kawahara, R. L. Lee, C. Tolman, J. G. Seidman, and E. J. Neer. 1988. Molecular characterization of cDNA encoding a new human GTP-binding protein. *Proceedings of the National Academy of Sciences.* 85:4153–4157.

Kurose, H., T. Katada, T. Haga, K. Haga, A. Ichiyama, and M. Ui. 1986. Functional interaction of purified muscarinic receptors with purified inhibitory guanine nucleotide regulatory proteins reconstituted in phospholipid vesicles. *Journal of Biological Chemistry.* 261:6423–6428.

Lerea, C. L., D. E. Somers, J. B. Hurley, I. B. Klock, and A. M. Bunt-Milam. 1986. Identification of specific transducin α subunits in retinal rod and cone photoreceptors. *Science.* 234:77–80.

Liang, B. T., E. J. Neer, and J. B. Galper. 1986. The non-coordinate development of muscarinic cholinergic inhibition and adrenergic stimulation of adenylate cyclase in embryonic chick heart. *Journal of Biological Chemistry.* 261:9011–9021.

Lochrie, M.A., and M. I. Simon. 1988. G-protein multiplicity in signal transduction systems. *Biochemistry.* 27:4957–4965.

Logothetis, D. E., D. Kim, J. K. Northup, E. J. Neer, and D. E. Clapham. 1988. Specificity of action of guanine nucleotide-binding regulatory protein subunits on the cardiac muscarinic K^+ channel. *Proceedings of the National Academy of Sciences.* 85:5814–5818.

McFadzean, I., I. Mullaney, D. A. Brown, and G. Milligan. 1989. Antibodies to the GTP binding protein, G_o, antagonize noradrenaline-induced calcium current inhibition in NG108-15 hybrid cells. *Neuron.* 3:177–182.

Mullaney, I., and G. Milligan. 1989. Elevated levels of the guanine nucleotide binding protein, G_o, are associated with differentiation of neuroblastoma x glioma hybrid cells. *FEBS Letters.* 244:113–118.

Navon, S. E., and B. K.-K. Fung. 1987. Characterization of transducin from bovine retinal rod outer segments. Participation of the amino-terminal region of T_α in subunit interaction. *Journal of Biological Chemistry.* 262:15746–15751.

Neer, E. J., and D. E. Clapham. 1988. Roles of G-proteins in transmembrane signalling. *Nature.* 333:129–134.

Neer, E. J., S.-Y. Kim, S. -L. Ang, D. B. Bloch, D. K. Bloch, Y. Kawahara, C. Tolman, R. Lee, D. Logothetis, D. Kim, J. G. Seidman, and D. E. Clapham. 1988a. Functions of G-protein subunits. *Cold Spring Harbor Symposia on Quantitative Biology.* 53:241–246.

Neer, E. J., L. Pulsifer, and L. G. Wolf. 1988b. The amino terminus of G protein α subunits is required for interaction with $\beta\gamma$. *Journal of Biological Chemistry.* 263:8996–9000.

Neer, E. J., J. M. Lok, and L. G. Wolf. 1984. Purification and properties of the inhibitory guanine nucleotide regulatory unit of brain adenylate cyclase. *Journal of Biological Chemistry.* 259:14222–14229.

Olate, J., H. Jorquera, P. Purcell, J. Codina, L. Birnbaumer, and J. E. Allende. 1989. Molecular cloning and sequence determination of a cDNA coding for the α-subunit of a G_o-type protein of *Xenopus laevis* oocytes. *FEBS Letters.* 244:188–192.

Peraldi, S., B. N. T. Dao, P. Brabet, V. Homburger, B. Rouot, M. Toutant, C. Bouille, I. Assenmacher, J. Bockaert, and J. Gabrion. 1989. Apical localization of the alpha subunit of GTP-binding protein G_o in choroidal and ciliated ependymocytes. *Journal of Neuroscience.* 9:806–814.

Price, S. R., S. -C. Tsai, R. Adamik, C. W. Angus, I. M. Serventi, M. Tsuchiya, J. Moss, and M. Vaughan. 1989. Expression of $G_{o\alpha}$ mRNA and protein in bovine tissues. *Biochemistry.* 28:3803–3807.

Schmidt, C. J., S. Garen-Fazio, Y. -K. Chow, and E. J. Neer. 1989. Neuronal expression of a newly identified *Drosophila melanogaster* G protein α_o subunit. *Cell Regulation.* 1:125–134.

Silbert, S., T. Michel, R. Lee, and E. J. Neer. 1990. Differential degradation rates of the G protein α_o in cultured cardiac and pituitary cells. *Journal of Biological Chemistry.* 265:3102–3105.

Sternweis, P. C. 1986. The purified α subunits of G_o and G_i from bovine brain require $\beta\gamma$ for association with phospholipid vesicles. *Journal of Biological Chemistry.* 261:631–637.

Sternweis, P. C., and J. Robishaw. 1984. Isolation of two proteins with high affinity for guanine nucleotides from membranes of bovine brain. *Journal of Biological Chemistry.* 259:13806–13813.

Thambi, N. C., F. Quan, W. J. Wolfgang, A. Spiegel, and M. Forte. 1989. Immunological and molecular characterization of $G_{o\alpha}$-like proteins in the *Drosophila* central nervous system. *Journal of Biological Chemistry.* 264:18552–18560.

Toutant, M., D. Aunis, J. Bockaert, V. Homburger, and B. Rouot. 1987. Presence of three pertussis toxin substrates and $G_o\alpha$ immunoreactivity in both plasma and granule membranes of chromaffin cells. *FEBS Letters.* 215:339–344.

Tsai, S. -C., R. Adamik, Y. Kanaho, J. L. Halpern, and J. Moss. 1987. Immunological and biochemical differentiation of guanyl nucleotide binding proteins: Interaction of $G_{o\alpha}$ with rhodopsin, anti-$G_{o\alpha}$ polyclonal antibodies, and a monoclonal antibody against transducin α subunit and $G_{i\alpha}$. *Biochemistry.* 26:4728–4733.

Ueda, H., H. Harada, M. Nozaki, T. Katada, M. Ui, M. Satoh, and H. Takagi. 1988. Reconstitution of rat brain μ opioid receptors with purified guanine nucleotide-binding regulatory proteins, G_i and G_o. *Proceedings of the National Academy of Sciences.* 85:7013–7017.

Ueda, H., Y. Yoshihara, H. Misawa, N. Fukushima, T. Katada, M. Ui, H. Takagi, and M. Satoh. 1989. The kyotorphin (tyrosine-arginine) receptor and a selective reconstitution with purified G_i, measured with GTPase and phospholipase C assays. *Journal of Biological Chemistry.* 264:3732–3741.

Van Dongen, A. M., J. Codina, J. Olate, R. Matera, R. Joho, L. Birnbaumer, and A. M. Brown. 1988. Newly identified brain potassium channels gated by the guanine nucleotide binding protein G_o *Science.* 242:1433–1437.

Van Dop, C., G. Yamanaka, F. Steinberg, R. D. Sekura, C. R. Manclark, L. Stryer, and H. R. Bourne. 1984. ADP-ribosylation of transducin by pertussis toxin blocks the light-stimulated hydrolysis of GTP and cGMP in retinal photoreceptors. *Journal of Biological Chemistry.* 259:23–26.

Van Meurs, K. P., C. W. Angus, S. Lavu, H. -F. Kung, S. K. Czarnecki, J. Moss, and M. Vaughan. 1987. Deduced amino acid sequence of bovine retinal $G_{o\alpha}$: similarities to other guanine nucleotide-binding proteins. *Proceedings of the National Academy of Sciences.* 84:3107–3111.

West, R. E., Jr., J. Moss, M. Vaughan, T. Kiu, and T. -Y. Liu. 1985. Pertussis toxin-catalyzed ADP-ribosylation of transducin. *Journal of Biological Chemistry.* 260:14428–14430.

Winslow, J. W., J. R. Van Amsterdam, and E. J. Neer. 1986. Conformations of the α_{39}, α_{41}, and $\beta.\gamma$ components of brain guanine nucleotide-binding proteins. *Journal of Biological Chemistry.* 261:7571–7579.

Worley, P. F., J. M. Baraban, C. Van Dop, E. J. Neer, and S. Snyder. 1986. G_o, a guanine nucleotide-binding protein: immunohistochemical localization in rat brain resembles distribution of second messenger systems. *Proceedings of the National Academy of Sciences.* 83:4561–4565.

Yatani, A., R. Mattera, J. Codina, R. Graf, K. Okabe, E. Padrell, R. Iyengar, A. M. Brown, and L. Birnbaumer. 1988. The G protein-gated atrial K^+ channel is stimulated by three distinct G_i α-subunits. *Nature.* 336:680–682.

Yoon, J., R. D. Shortridge, B. T. Bloomquist, S. Schneuwly, M. H. Perdew, and W. L. Pak. 1989. Molecular characterization of *Drosophila* gene encoding G_o α subunit homolog. *Journal of Biological Chemistry.* 264:18536–18543.

Chapter 13

The Molecular Components of Olfaction

Randall R. Reed

Howard Hughes Medical Institute, Department of Molecular Biology and Genetics, The Johns Hopkins Medical School, Baltimore, Maryland 21205

The mammalian olfactory system is an exquisitely sensitive sensory organ responsible for encoding information on both the intensity and the identity of a wide variety of chemical stimuli. The posterior portion of the nasal cavity consists of the olfactory neuroepithelium responsible for the conversion of the odorant stimulus into an electrical signal. The cell bodies of the sensory neurons reside in this neuroepithelium and extend a single dendritic process to the luminal surface where they terminate in a structure referred to as the dendritic knob. A small number of very thin, nonmotile cilia extend from the dendritic knob into the mucous layer. The cilia and the dendritic tip are the only part of the sensory neuron exposed to the external environment and are therefore thought to be the site of transmembrane signaling. Each of the sensory neuron cell bodies, located in the middle portion of the epithelium, project a single unbranched axon through the cribriform plate and synapse on their target cells in the glomerular tufts of the olfactory bulb. The olfactory neurons have the unique capacity to be continually replaced from a population of neuroblast precursors throughout adult life. More significantly, acute injury to the olfactory bulb, the olfactory nerve, or the receptor neurons themselves, leads to the rapid loss of the mature sensory cells from the epithelium and their subsequent, synchronous replacement (Jones et al., 1988).

The mechanisms underlying the perception of odorants has been studied by a variety of electrophysiological, biochemical, genetic, and psychophysical approaches. For example, single-cell recordings have suggested that only a fraction of the sensory neurons in the epithelium respond to any particular odorant. This has led to models for odorant discrimination which use receptor specificity and the differential electrical responses of individual cells to transmit information on chemical stimulus identity to the brain. The specific receptor proteins could therefore converge on a common intracellular pathway. Previous biochemical evidence suggested that in partially purified sensory cilia at least some odorants could stimulate adenylate cyclase activity (Pace et al., 1985). A central component of many of the known second messenger systems is the GTP-binding protein. The guanine-binding proteins (G proteins) are heterotrimers of $\alpha\beta\gamma$ subunits that couple membrane-bound receptors to second messenger enzymes or ion channels. The α subunit appears to confer identity to the multimer and in most systems governs the specificity of the interaction with receptors and effectors. We have undertaken experiments designed to identify the components of the olfactory signal transduction cascade. The G protein α subunit plays a central role in these pathways and we have focused our initial efforts on attempts to identify the nature of the G protein involved in olfaction.

The considerable homology shared among the G_α subunits at both the protein and nucleotide sequence level allowed us to identify cDNAs that encode each of the abundant species present in olfactory neuroepithelium (Jones and Reed, 1987). Of the six distinct classes of cDNAs that were identified, one is expressed exclusively and abundantly in RNA isolated from olfactory tissue (Jones and Reed, 1989). We have determined which cells within the olfactory neuroepithelium express this novel G protein that we have termed G_{olf} by examining the level of expression of G_{olf} mRNA in normal olfactory tissue or alternatively after depletion of the sensory neurons from the epithelium. The mRNA that encodes G_{olf} undergoes a dramatic decrease when the neurons are depleted. This change in the level of expression is paralleled by a similar decrease in a known olfactory neuron specific protein, OMP (olfactory marker protein).

Odorants appear to stimulate adenylyl cyclase in a GTP-dependent manner in olfactory cilia preparations. If G_{olf} is mediating olfactory transduction, its predicted protein sequence might be most similar to the known α subunit, G_s, which interacts with adenylyl cyclase in other systems. The sequence of the G_{olf_α} cDNA reveals a striking, 88% amino acid identity with the G_{s_α} subunit.

To obtain direct biochemical evidence consistent with a role for G_{olf} in mediating olfaction, we have used a retrovirus expression system to introduce the coding region for G_{olf} into S49 cells. The S49 mouse lymphoma cell line is deficient in GTP-stimulated adenylyl cyclase activity and has proven to be a useful system to investigate G protein function. When G_{olf} is introduced into this cell line, the activation of G_{olf} protein by nonhydrolyzable GTP analogues leads to the stimulation of adenylyl cyclase activity. G_{olf} is therefore competent to mediate the effector function thought to be responsible for olfactory signal transduction. More significant is the observation that β-adrenergic agonists can activate adenylyl cyclase in S49 cells that have been transfected with the G_{olf} expression vector (Jones et al., 1990). These data imply that a G protein–coupled receptor that normally interacts with G_s, the β-adrenergic receptor, can stimulate GTP exchange on G_{olf} and lead to the activation of the effector enzyme. Given the purported role for G_{olf} in olfaction, it seems likely that olfactory receptors may be structurally similar to the β-adrenergic receptors which have been extensively studied at the molecular level.

The mammalian olfactory system appears to have evolved a novel GTP-binding protein α subunit to mediate olfaction. We were interested whether the other components of the cascade had evolved similar olfactory-specific counterparts. Clearly, the odorant binding activity required of the olfactory receptors suggests that they will be specialized proteins expressed only by the olfactory sensory neurons. The catalytic activity of the effector enzyme, adenylyl cyclase, exists in essentially all cells. Recently, our laboratory in collaboration with Dr. Alfred Gilman at the University of Texas at Dallas, have identified and characterized cDNA clones that encode this integral membrane protein (Krupinski et al., 1989). Using low stringency hybridization, there appears to be a novel form of this enzyme expressed by olfactory neurons. The special kinetic properties of this enzyme, if any, and its role in olfaction remain to be determined. Presently, we are also attempting to identify olfactory receptors through a variety of molecular cloning approaches. These include exploiting similarities among the known members of the G protein–coupled receptor family.

References

Jones, D. T., E. Barbosa, and R. R. Reed. 1988. Expression of G-protein α subunits in rat olfactory neuroepithelium: candidates for olfactory signal transduction. *Cold Spring Harbor Symposium on Quantitative Biology.* 53:349–353.

Jones, D. T., S. B. Masters, H. R. Bourne, and R. R. Reed. 1990. Biochemical characteristics of three stimulatory GTP-binding proteins: the large and small forms of G_s and the olfactory specific G-protein, G_{olf}. *Journal of Biological Chemistry.* 265:2671–2676.

Jones, D. T., and R. R. Reed. 1987. Molecular cloning of five GTP-binding protein cDNA species from rat olfactory epithelium. *Journal of Biological Chemistry.* 262:14241–14249.

Jones, D. T., and R. R. Reed. 1989. G_{olf}: an olfactory neuron specific G-protein involved in odorant signal transduction. *Science.* 244:730–736.

Krupinski, J., F. Coussen, H. A. Bakalyar, W. J. Tang, P. G. Feinstein, K. Orth, C. Slaughter, R. R. Reed, and A. G. Gilman. 1989. Amino acid sequence of an adenylyl cyclase suggests a channel- or transporter-like structure. *Science*. 244:1558–1564.

Pace, U., E. Hanski, Y. Salomon, and D. Lancet. 1985. Odorant-sensitive adenylate cyclase may mediate olfactory reception. *Nature*. 316:255–258.

Chapter 14

Drosophila G Proteins

James B. Hurley, Stuart Yarfitz, Nicole Provost, and Sunita de Sousa

Howard Hughes Medical Institute and Department of Biochemistry, University of Washington, Seattle, Washington 98195

Introduction

G proteins mediate a variety of physiological processes ranging from phototransduction to yeast-mating responses (Gilman, 1987). The basic mechanism of G protein action begins when a stimulated receptor catalyzes GTP binding to the G protein α subunit. The GTP charged α subunit then alters the activity of an effector. Phosphodiesterases, cyclases, and ion channels have all been identified as G protein–coupled effectors.

Several activities contribute to G protein function. A G protein binds to a stimulated receptor with a certain affinity. In this complex, exchange of previously bound GDP for GTP occurs at a certain rate. When a G protein α subunit binds GTP the affinity of the α subunit for the $\beta\gamma$ complex weakens. The dissociated α subunit–GTP complex then interacts with the target effector with a certain specificity and affinity. This interaction persists until the α subunit hydrolyzes the bound GTP to GDP at a specific rate. Each of these activities could potentially determine the kinetics or sensitivity of a G protein–mediated response.

Examples of G protein–mediated responses with quite different sensitivities and kinetics are rod and cone phototransduction (Baylor, 1987). Rod cells are exquisitely sensitive to light but slow to respond whereas cones are much less sensitive to light but quick. Rod and cone cells each express different photoreceptor pigments, G proteins, and phosphodiesterases (Hurwitz et al., 1984; Lerea et al., 1986; Nathans et al., 1986). Different biochemical properties of any one or all of these signal transduction enzymes could be responsible for the dramatically different physiological responses of these cells.

To investigate how the kinetic properties of G proteins determine the speed and sensitivity of physiological responses we are studying G proteins in organisms with which genetic manipulations are possible. In one such study the effects of expressing mutant G proteins in *Drosophila melanogaster* are being investigated. In order to perform such an analysis we first characterized several *Drosophila* G proteins and the genes that encode them. These three *Drosophila* G protein genes are described here.

Materials and Methods

The methods used to screen *Drosophila* genomic (Maniatis et al., 1978) and cDNA libraries (Poole et al., 1985; provided by P. Salvaterra, Beckman Institute, City of Hope, CA) are described in Provost et al. (1988), Yarfitz et al. (1988), and de Sousa et al. (1989).

In situ hybridization analyses on paraffin sections of *Drosophila* tissue were performed as described by de Sousa et al. (1989) using antisense riboprobes derived from DGα1, DGα2, or DGβ cDNA clones. In each case, sense probes derived from the same cDNAs were used as controls and showed no detectable specific hybridization to the tissue sections.

DGα1 antipeptide antibodies were generated in rabbits using the synthetic peptide LDRIAQPNYIPTQQD coupled to either rabbit or bovine serum albumin as antigen. The antibodies were affinity purified on a peptide-sepharose column as described elsewhere (Lerea et al., 1986). TrpE-DGα1 fusion proteins were made using a Path 3 vector fused to a 3' Eco R1 cDNA fragment of the DGα1 cDNA. The 58-kD product synthesized in bacteria was partially purified by electroelution from sodium dodecyl sulfate (SDS) polyacrylamide gels before being injected into rabbits for antibody production. Affinity-purified "LD" antibody (Goldsmith et al., 1987) was obtained from Dr. Allen Spiegel (National Institutes of Health).

Monoclonal antibodies against the *Drosophila* β subunit were isolated by injecting mice with a TrpE-DGβ fusion protein containing amino acids 1–340 of DGβ. Hybridomas were selected that produced antibodies that bound to the fusion protein, but not to TrpE protein itself.

Immunocytochemical analyses were performed on frozen sections of *Drosophila* tissue from flies that were embedded in OCT (Miles Laboratories Inc., Naperville, CT) frozen, cut into 10-μm sections, and fixed in 4% paraformaldehyde. The sections were then incubated with primary antibodies, which were then visualized using biotinylated secondary antibodies coupled to streptavidin-horseradish peroxidase conjugate and a solution of diaminobenzedine.

Cellular components of adult *Drosophila* were fractionated by a variation of the method of Wu (1988). Proteins were isolated from these fractions and analyzed by immunoblotting as described by Lerea et al., 1986.

Results

Isolation of *Drosophila* G Protein Genes

Several observations have previously demonstrated the presence of G proteins in *Drosophila*. (*a*) [^{32}P]-azido GTP labels several ~40 kD proteins in insect eye homogenates. One of these proteins appears to be involved with phototransduction since its association with [^{32}P]-azido GTP is light dependent (Devary et al., 1987). (*b*) Pertussis toxin, an ADP-ribosyl transferase that modifies only certain G protein α subunits, recognizes a 39-kD protein from *Drosophila* homogenates (Hopkins et al., 1988). (*c*) GTP is required for rhodopsin-mediated light activation of phospholipase C in extracts of *Musca* eyes (Devary et al., 1987).

Genes encoding *Drosophila* G proteins were isolated by screening a *Drosophila* genomic DNA library at reduced stringency with mammalian G protein cDNAs (Provost et al., 1988; Yarfitz et al., 1988; de Sousa et al., 1989). Genes corresponding to Gαi, Gαo, and Gβ were isolated in this way. These genes are referred to as DGα1, DGα2, and DGβ, respectively. DGα1 maps to position 65C on the *Drosophila* third chromosome by in situ hybridization to polytene chromosomes. DGα2 maps to 47A on the second chromosome and DGβ to 13F on the X chromosome. None of these map positions correspond to loci of any reported *Drosophila* mutants.

The proteins encoded by these genes are very similar to their mammalian counterparts. DGα1 is 78% identical to mammalian Gαi1, DGα2 is 83% identical to mammalian Gαo, and DGβ is 83% identical to mammalian Gβ. Two differently spliced mRNA transcripts of DGα2 were identified. These two transcripts are derived from the same gene, but appear to be produced by two different promoters (Yoon et al., 1989). If both are translated, they would generate proteins differing only at seven amino acids near their amino termini. Both forms of DGα2 protein would have a CAAX sequence at their carboxy termini. This sequence is recognized as a site for ADP-ribosylation by pertussis toxin. DGα1 lacks this recognition site. Therefore, DGα2 is most likely to be the pertussis toxin substrate identified by Hopkins et al. (1988).

Multiple DGβ genes are present in mammalian genomes (Fong et al., 1987). However, there appears to be only one β subunit gene in *Drosophila* based on low stringency genomic Southern blot analysis (Yarfitz et al., 1988).

Expression of *Drosophila* G Protein Genes

To obtain information about the physiological importance of each of these genes, we examined their spatial and developmental expression by in situ hybridization. DGα1, DGα2, and DGβ cDNA clones were isolated and subcloned into vectors that have T3 and T7 phage promoters bracketing the cDNA clone. The phage promoters were used to synthesize antisense and sense ^{35}S-labeled RNA probes to identify RNA transcripts derived from the *Drosophila* G protein genes. Fig. 1 shows that DGβ transcripts are present in cell bodies of the adult brain as well as throughout the *Drosophila* nervous system. In addition, DGβ transcripts are present in ovaries, particularly in nurse cells and oocytes at advanced stages of development (not shown). Only low levels of DGβ mRNA are detected in *Drosophila* eyes. DGβ message is far less abundant in eyes than in the cell bodies of other neural tissue such as the brain and thoracic ganglion.

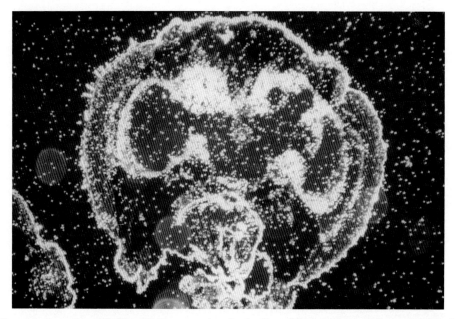

Figure 1. Location of DGβ mRNA transcripts in the *Drosophila* adult head determined by in situ hybridization. See de Sousa et al. (1989) for identification of tissues in *Drosophila* heads.

DGα2 transcripts are also widely expressed throughout the nervous system (Fig. 2). However, no transcripts are detected in *Drosophila* eyes. DGα2 is also present in the adult ovary and in the developing nervous system of late stage embryos (not shown). Probes specific for the two different DGα2 transcripts revealed similar expression in the nervous system. However, a difference in expression was detected in ovaries (de Sousa et al., 1989).

The pattern of DGα1 expression is unique and unexpected (Fig. 3). Little DGα1 message is detected in the adult nervous system. However, small amounts are present in pupal lamina. DGα1 transcripts are abundant only in early embryos, in nurse cells, and in oocytes at advanced stages of development within the ovary. These results agree with a previous developmental analysis of DGα1 expression in which RNA preparations from different stages of *Drosophila* development were analyzed by Northern blot analysis (Provost et al., 1988).

Proteins encoded by these genes were located by producing antibodies that recognize the DGβ and DGα1 gene products. Mouse monoclonal DGβ antibodies were raised against a bacterially expressed fusion protein with an amino terminus encoded by the bacterial trpE gene and the carboxy terminus derived from the full coding region of the DGβ cDNA. Two hybridomas were selected that produced antibodies that consistently showed positive immunoreactivity on sections of *Drosophila* tissue. These antibodies recognize the entire adult *Drosophila* nervous system including brain, thoracic ganglion, and rhabdomeres within the *Drosophila* eye (Fig. 4). On immunoblots, these monoclonal antibodies also recognize a 37-kD protein in *Drosophila* homogenates and a 36-kD protein in bovine transducin.

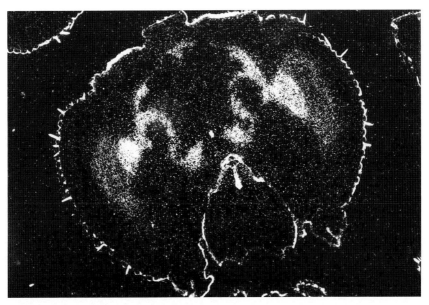

Figure 2. Location of DGα2 mRNA transcripts in the *Drosophila* head determined by in situ hybridization.

Several rabbit polyclonal antibodies that recognize the DGα1 gene product were also produced. One antibody was prepared in rabbits by injecting rabbit serum albumin conjugated to the peptide LDRIAQPNYIPTQQD. This identical sequence is also present in bovine Gαi1. Therefore, a previously characterized anti-peptide antibody raised against the peptide LDRIAQPNYI was also obtained from Dr. Alan Spiegel and his colleagues (Goldsmith et al., 1988). A third type of antibody to DGα1 was also produced by injecting into rabbits a bacterially expressed fusion protein with its amino terminus encoded by the bacterial trpE gene and the carboxy terminus made up of amino acids 159–355 of the DGα1 protein. Each antibody recognized the same tissues in frozen sections of *Drosophila* (Fig. 5 A). Consistent with the pattern of DGα1 RNA expression, little or no DGα1 protein was detected in adult nervous tissue. However, DGα1 protein was detected within ovaries, primarily within nuclei of nurse

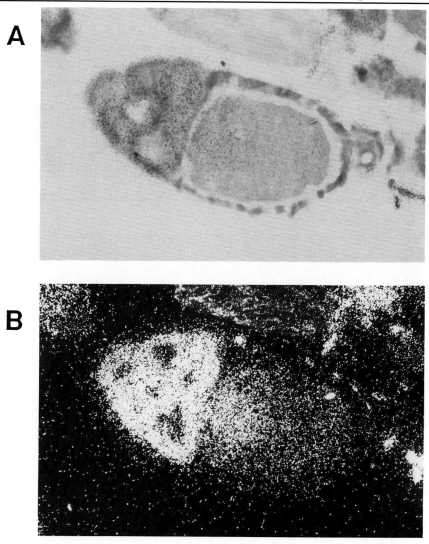

Figure 3. Expression of DGα1 mRNA transcripts in a developing ovariole. (*A*) Bright field showing Giemsa stained ovarian chamber. (*B*) Dark field showing that transcripts are abundant in nurse cells and are being dumped into the oocyte at this stage of development. See de Sousa et al. (1989) for identification of tissues in *Drosophila* ovaries.

cells and follicle cells. The strong nuclear signal was not present when the anti-peptide antibodies were preincubated with a 100× excess of the LDRIAQPNYIPTQQD peptide. (Fig. 5 *B*). A 1,000× excess of a related, but different peptide derived from rod transducin did not block the signal (not shown). The location of DGβ protein in ovaries was also examined using the DGβ monoclonal antibodies described above. Although most DGβ protein is present throughout neural tissue, there is also strong DGβ immunoreactivity in the nurse cell and follicle cell nuclei (not shown).

To confirm the location of DGα1, nuclei were isolated from homogenates of

Drosophila abdomens by a series of low speed spins (Wu, 1988). Nuclear and nonnuclear proteins were analyzed by SDS polyacrylamide gel electrophoresis and immunoblotting using the antipeptide and antifusion protein antibodies. The antipeptide antibody ("LD") prepared by Goldsmith et al. (1988) described above recognizes a 38-kD protein present only in the nuclear fraction (Fig. 6). This signal is not detected when the antibody is preincubated with a 100× molar excess of the appropriate peptide. The same fractions were also probed with one of the DGβ monoclonal antibodies. In striking contrast to the distribution of DGα1 protein, most of the 37-kD DGβ protein is present in the nonnuclear fraction (not shown).

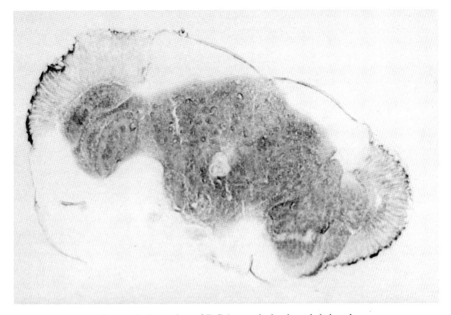

Figure 4. Location of DGβ protein in the adult head.

Discussion

G proteins in organisms that diverged as long ago as yeast and mammals are similar in structure which suggests that they are derived from a common ancestral gene and that many of their functions have been conserved throughout evolution. Here we demonstrate that *Drosophila melanogaster* has its complement of G protein genes as well. Previously, G proteins in *Drosophila* had been identified using bacterial toxins that specifically ADP-ribosylate G protein α subunits. In addition, G proteins have been implicated in invertebrate phototransduction (Blumenfeld et al., 1985; Devary et al., 1987). The expression patterns of three *Drosophila* G protein subunit genes are described in this report. A fourth G protein subunit gene, encoding a G_s-like α subunit, has also been characterized (Quan et al., 1989).

Two of the *Drosophila* G proteins investigated in this report appear to be expressed at high levels throughout the nervous system. DGα2 mRNA and protein are

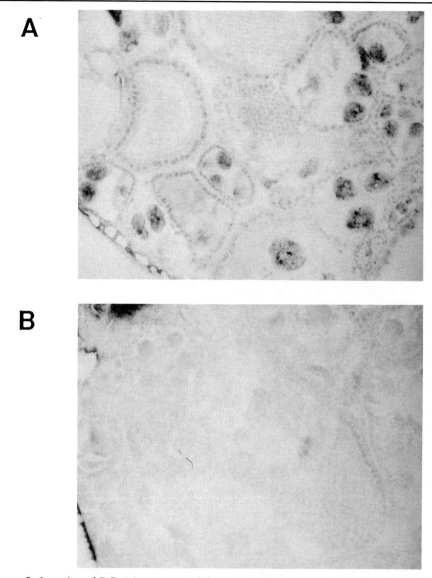

Figure 5. Location of DGα1 immunoreactivity in *Drosophila* ovary. (*A*) Section probed with affinity-purified antibody. (*B*) Adjacent section probed with the same preparation of antibody that had been preincubated with 100× molar excess of the peptide.

present in the brain and thoracic ganglion based on in situ hybridization results and pertussis toxin labeling. DGβ immunoreactivity is found throughout the nervous system including eyes. The presence of DGβ protein in eyes was unexpected because only small amounts of DGβ mRNA transcripts were detected there. Either the transcripts are translated more efficiently in eyes than in brain, the protein is more stable in eyes, or the antibody is detecting another form of DGβ not recognized by the cDNA probe. DGα1 has the most unexpected pattern of expression. The DGα1 protein is present mainly in nuclei of nurse cells and follicle cells of ovaries. None of the α subunits characterized either in this report or elsewhere are present in the eye, so the identity of a G protein α subunit that could participate in phototransduction remains uncertain.

It is not clear whether or not the three *Drosophila* G protein genes, DGα1 or DGα2 reported here or the G_s α subunit reported elsewhere (Quan et al., 1989) represent the entire complement of *Drosophila* G protein α subunit genes. Several labs have now repeatedly isolated the same three cDNA and genomic α subunit clones. @owever, all three laboratories have screened the same cDNA and genomic libraries (de Sousa et al., 1989; Quan et al., 1989; Yoon et al., 1989). Further screening of other libraries and use of polymerase chain reaction methods to produce G protein enriched

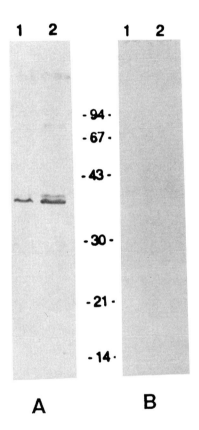

Figure 6. Immunoblot analysis of nuclei prepared from *Drosophila* abdomens. (*A*) 15 μg of total protein from a homogenate of *Drosophila* abdomens were loaded in lanes *1* and *15* μg of protein from a homogenate of isolated nuclei were loaded in lane *2*. Blot was probed with affinity-purified antipeptide antibody (Goldsmith et al., 1988) at 0.3 μg/ml. (*B*) Identical blot probed with antibody preincubated with a 100× molar excess of LDRIAQPNYIPTQQD peptide.

libraries will be necessary to determine if other G protein genes reside in the *Drosophila* genome.

Identification of the physiological roles of the G protein subunits described here will depend on expression studies. Recently, mutations that block GTP hydrolysis activity have been identified in G protein α subunits (Graziano and Gilman, 1989; Masters et al., 1989). Expression of G protein subunits with these dominant mutations, coupled with the localization studies reported here should help to identify the roles that G proteins and their kinetic properties play in development and in physiological responses to stimuli.

Acknowledgments

We thank Glenda Froelick for preparation of paraffin and frozen sections and Erik Jorgensen for advice on in situ hybridization techniques. We are grateful to Allen Spiegel and his colleagues for supplying affinity-purified LD antibody.

References

Baylor, D. A. 1987. Photoreceptor signals and vision. *Investigations in Opthalmology and Visual Science*. 28:34–49.

Blumenfeld, A., J. Erusalinsky, O. Heichal, Z. Selinger, and B. Minke. 1985. Light activated guanosinetriphosphatase in *Musca* eye membranes resembles the prolonged depolarizing afterpotential in photoreceptor cells. *Proceedings of the National Academy of Sciences*. 82:7116–7120.

de Sousa, S. M., L. L. Hoveland, S. Yarfitz, and J. B. Hurley. 1989. The *Drosophila* $G_o\alpha$-like protein gene produces multiple transcripts and is expressed in the nervous system and in ovaries. *Journal of Biological Chemistry*. 264:18544–18551.

Devary, O., O. Heichal, A. Blumenfeld, D. Cassel, E. Suss, S. Barash, C. T. Rubinstein, B. Minke, and Z. Selinger. 1987. Coupling of photoexcited rhodopsin to inositol phospholipid hydrolysis in fly photoreceptors. *Proceedings of the National Academy of Sciences*. 84:6939–6943.

Fong, H. K.-W., T. T. Armatruda, B. W. Birren, and M. I. Simon. 1987. Distinct forms of the β subunit of GTP-binding regulatory proteins indentified by molecular cloning. *Proceedings of the National Academy of Sciences*. 84:3792–3796.

Gilman, A. G. 1987. G-proteins: transducers of receptor-generated signals. *Annual Review of Biochemistry*. 56:615–649.

Goldsmith, P., P. Gierschik, G. Mulligan, C. G. Unson, R. Vitinsky, H. L. Malech, and A. M. Spiegel. 1988. Antibodies directed against synthetic peptides distinguish between GTP-binding proteins in neutrophil and brain. *Journal of Biological Chemistry*. 263:6476–6479.

Graziano, M. P., and A. G. Gilman. 1989. Snythesis in Escherichia coli of GTPase deficient mutants of $G_s\alpha$. *Journal of Biological Chemistry*. 264:15475–15482.

Hopkins, R. S., M. A. Stamnes, M. I. Simon, and J. B. Hurley. 1988. Cholera toxin and pertussis toxin substrates and endogenous ADP-ribosylation activity in *Drosophila melanogaster*. *Biochimica et Biophysica Acta*. 970:355–362.

Hurwitz, R. L., A. H. Bunt-Milam, and J. A. Beavo. 1984. Immunologic characterization of the photoreceptor outer segment cylic GMP phosphodiesterase. *Journal of Biological Chemistry*. 259:8612–8618.

Lerea, C. L., D. E. Somers, J. B. Hurley, I. B. Klock, and A. H. Bunt-Milam. 1986. Isolation of specific transducin α subunits in retinal rod and cone photoreceptors. *Science*. 234:77–80.

Maniatis, T., R. C. Hardison, E. Lacy, J. Lauer, C. O'Connell, D. Quon, G. K. Sim, and A. Efstratiadis. 1978. The isolation of structural genes from libraries of eucaryotic DNA. *Cell*. 15:687–701.

Masters, S. B., R. T. Miller, M.-H. Chi, F.-H. Chang, B. Beiderman, N. G. Lopez, and H. R. Bourne. 1989. Mutations in the GTP-binding site of $G_s\alpha$ alterstimulation of adenylyl cyclase. *Journal of Biological Chemistry*. 264:15467–15474.

Nathans, J., D. Thomas, and D. S. Hogness. 1986. Molecular genetics of human color vision: the genes encoding blue, green and red pigments. *Science*. 232:193–202.

Poole, S. J., L. M. Kauvar, B. Drees, and T. Kornberg. 1985. The engrailed locus of *Drosophila:* structural analysis of an embryonic transcript. *Cell.* 40:37–43.

Provost, N. M., D. E. Somers, and J. B. Hurley. 1988. A *Drosophila melanogaster* G-protein α subunit gene is expressed primarily in embryos and pupae. *Journal of Biological Chemistry.* 263:12070–12076.

Quan, F., W. Wolfgang, and M. Forte. 1989. The *Drosophila* gene coding for the α subunit of a stimulatory G-protein is preferentially expressed in the nervous system. *Proceedings of the National Academy of Sciences.* 86:4321–4325.

Wu, C. 1988. Analysis of hypersensitive sites in chromatin. *Methods in Enzymology.* 170:269–284.

Yarfitz, S., N. M. Provost, and J. B. Hurley. 1988. Cloning a *Drosophila melanogaster* guanine nucleotide regulatory protein β subunit gene and characterization of its expression during development. *Proceedings of the National Academy of Sciences.* 85:7134–7138.

Yoon, J., R. D. Shortridge, B. T. Bloomquist, S. Schnewly, M. H. Perdew, and W. L. Pak. 1989. Molecular characterization of a *Drosophila* gene encoding $G_o\alpha$ subunit homolog. *Journal of Biological Chemistry.* 264:18536–18543.

Chapter 15

G Protein Coupling of Receptors to Ionic Channels and Other Effector Systems

Lutz Birnbaumer, Atsuko Yatani, Antonius M. J. VanDongen, Rolf Graf, Juan Codina, Koji Okabe, Rafael Mattera, and Arthur M. Brown

Departments of Cell Biology and Molecular Physiology and Biophysics, Baylor College of Medicine, Houston, Texas 77030

Introduction

G proteins play a central role in coupling receptors to effector systems. Like receptors, which are increasing in number rapidly, effectors affected by the activated forms of G proteins are also increasing, most notably through the discovery in 1986–1987 that ionic channels form part of the family of molecules regulated by G proteins. Using cell-free systems such as those provided by excision of membrane patches from cells and incorporation of plasma membrane vesicles into lipid bilayers, it was shown that ionic channels are indeed regulated by activated G proteins. Some of these channels had long before been predicted to be under the control of G protein–coupled receptors by means other than soluble second messengers. The mechanism by which G proteins regulate some of these channels is at the very least "membrane delimited" and independent of any phosphorylation event or of changes in cytoplasmic levels of second messengers such as cAMP, Ca^{2+}, or IP_3 and is very likely due to direct interaction of the G protein α subunit and the channel proper (for review see Brown and Birnbaumer, 1988). Our group was prominent in providing some of the initial as well as subsequent supporting data for these conclusions (Brown and Birnbaumer, 1988; Yatani et al., 1987a–c, 1988a–c; Codina et al., 1987a, b, 1988; Imoto et al., 1988; Kirsch et al., 1988). The recognition that G proteins may affect ionic channels under cell-free conditions led us as well as others to investigate this possibility further. By the most recent count (August, 1989) six classes of ionic channels comprising at least 12 separate molecular species defined by distinct kinetic and pharmacological properties, have been shown to be stimulated or inhibited under cell-free conditions by exogenously added G protein α subunits or by activation of a nearby G protein by GTPγS, as seen in inside-out membrane patches or after incorporation of membrane vesicles into planar lipid bilayers (summarized in Table I). These ionic channels include channels that are essentially silent unless a G protein is activating them, referred to as G protein–gated ion channels (Yatani et al., 1987a, c; VanDongen et al., 1988), as well as channels that are merely regulated by the G proteins, such as ATP-sensitive channels (Parent and Coronado 1989; Ribalet et al., 1989; Kirsch, G., L. Birnbaumer, and A. M. Brown, unpublished), and amiloride-sensitive channel (Light et al., 1989), a Ca^{2+}-activated charybdotoxin-sensitive channel (Toro et al., 1990), and various voltage-gated channels (Yatani et al., 1987b, 1988a, Dolphin et al., 1988; Schubert et al., 1989) (summarized in Table I).

In addition to the ~12 channels thus far shown to be influenced by G proteins in a manner that appears to be direct, there may still be several more. This is suggested by reports on effects of pertussis toxin (PTX) treatment or of GTPγS and G protein subunit injection on whole-cell currents. These effects include the inhibition by GTPγS or PTX-sensitive G_o- and G_i-type G proteins of voltage-gated Ca^{2+} channels in chick (Holz et al., 1986) and rat (Scott and Dolphin, 1986; Ewald et al., 1988, 1989) dorsal root ganglion cells and of similar channels in AtT-20 (Lewis et al., 1986) and NG108-15 (Hescheler et al., 1987) cells and the stimulation by PTX-sensitive G_i-type G protein of voltage-gated Ca^{2+} channels in adrenal Y1 and GH_3 cells (Hescheler et al., 1988; Rosenthal et al., 1988; summarized in Table II). However, in these cases the involvement of soluble second messengers in the mediation of the effects of the G protein has yet to be ruled out.

TABLE I
Ionic Channels Regulated Under Cell-free Conditions by Pure G Protein and/or GTPγS

Channel type	G protein	Effect	Tissue/cell
K_{Gk} 40 pS	G_i (1, 2, 3)	Stim.	Heart
K_{Gk} 50 pS	G_{i3}	Stim.	GH_3 cells
K_{Gk} 4 types	G_{o1}	Stim.	Hippocampal neurons
K_{ATP}	G_{i3}	Stim.	RIN cells, heart, skeletal muscle
$Na/K_{[Amil]}$	G_{i3}	Stim.	Renal medullary collecting tubule
$Ca_{[DHP]}$ L-type	G_s (all 4)	Stim.	Heart, skeletal muscle
$Na_{[TTX]}$	G_s	Inh.	Heart
$K_{Ca[Charyb]}$	? (Iso/GTP)	Stim.	Uterine smooth muscle
$Ca_{[Ni]}$ T-type	? (GTPγS)	Inh.	Rat dorsal root ganglions

The case of cell-free regulation of dihydropyridine (DHP)-sensitive Ca^{2+} channels by G_s is of special interest. It was unexpected for two reasons: first because DHP-sensitive Ca^{2+} channels had also been shown to be stimulated upon phosphorylation by the catalytic unit of cAMP-dependent protein kinase (Kameyama et al., 1986), indicating that nature uses dual pathways to regulate a single function, one fast and membrane delimited, the other slower with a longer life span; second, because it had been thought that G proteins might be "monogamous," i.e., specific for single effector functions, and here we were faced with proof for multifunctionality in G protein actions.

These advances were all the result of a multidisciplinary approach to the problem of signal transduction by G proteins which brought together classical biochemistry, sophisticated single-channel recordings and modern molecular biology. The background experiments, especially those of Nargeot et al. (1983), Soejima and Noma (1984), Pfaffinger et al. (1985), and Breitwieser and Szabo (1985), which led to the

TABLE II
Ionic Channels That May Be under Direct Regulation by G Proteins as Inferred from Whole-Cell Recordings

Channel type	Agonist	G protein	Effect	Cell	Mimicked by TPA
Ca (N-type?)	Opioid	$G_o > G_i$	Inh.	NG108-15	?
Ca (L-type)	Ang II	G_i	Stim.	Y1 adrenal	?
Ca (L-type)	SST	? GTPγS	Inh.	AtT-20	Yes
Ca (L-type)	GnRH	G_i/G_o	Stim.	GH_3	?
Ca (L-type)	SST	G_i/G_o	Inh.	GH_3	?
Ca (N-type?)	$GABA_{(B)}$	? GTPγS	Inh.	Rat DRG	No
Ca (?)	NPY	$G_o > G_i$	Inh.	Rat DRG	Yes
	BDK	$G_o = G_i$	Inh.	Rat DRG	Yes
Ca (?)	$GABA_{(B)}$	? GTPγS	Inh.	Chick DRG	Yes
	NE (αAR)	? GTPγS	Inh.	Chick DRG	Yes

All the G proteins involved are PTX sensitive. Ang, angiotensin; BDK; bradykinin; DRG, dorsal root ganglion; GABA; γ-aminobutyric acid; NE, norepinephrine; NPY, neuropeptide Y; SST, somatostatin; TPA, 12-O-tetradecanoyl-phorbol-13-acetate.

discovery that ionic channels are effector systems of G proteins akin to adenylyl cyclase and cGMP-phosphodiesterase, and the initial experiments showing effects of purified G proteins and protein subunits on channels in excised membrane patches were reviewed in Brown and Birnbaumer (1988). The present article will center on some of the more recent results from our laboratories dealing with a PTX-sensitive G_i-type family of G proteins, their effects on ionic channels, our efforts in assigning defined functions to individual G proteins as they are known from biochemical and molecular cloning studies, and some speculations that follow from the results obtained as to why G proteins dissociate and how transduction pathways are set up.

Primary Structure of G Proteins

The primary $\alpha\beta\gamma$ structure of G proteins has been reviewed recently by Lochrie and Simon (1988), as well as by us (Birnbaumer et al., 1989, 1990). The keys to understanding their functioning requires an understanding of both their subunit

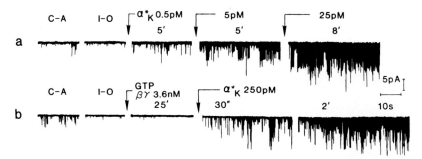

Figure 1. (*a*) Stimulation of single-channel K^+ currents in GH_3 cell membrane patches by increasing concentrations of native GTPγS-activated human erythrocyte α_{i3} (α_k^*). (*b*) Lack of intrinsic stimulatory effects of human erythrocyte $\beta\gamma$ dimer in a responsive membrane patch. 500 nM Lubrol PX was present throughout. (Adapted from Codina et al., 1987a.)

dissociation cycle superimposed on their GTPase cycle, and of their transient - or perhaps not so transient - interactions with receptors and effectors. In our view it is clear that the receptor signal is carried to the effector by the α subunit as exemplified in the experiment shown Fig. 1. At the time of this writing 12 α subunits, encoded in nine genes, two β subunits, and at least two γ subunits are known. Of these, all the α, the two β, and one of the γ subunits have been cloned. β and γ subunits form dimers which may be of two types, $\beta_{36}\gamma$ and $\beta_{35}\gamma$, if a cell expresses only one type of γ subunit, or of four types if it expresses two. To our knowledge, α subunits, when combining with $\beta\gamma$ dimers to form holo-G proteins, do not distinguish among $\beta\gamma$ dimers (Okabe et al., 1990). This is not to say that all tissues have the same complement of $\beta\gamma$ dimers. Quite the contrary, as has been shown recently the β_{35}/β_{36} ration differs in human placenta, human erythrocytes, bovine brain, and bovine retinal rod cells (Okabe et al., 1990). α subunits bind GTP, hydrolyze GTP, dissociate from the $\beta\gamma$ dimer on activation by GTP analogues and or the $NaAlF_4$/GDP, and with few exceptions are substrates for ADP-ribosylation by cholera toxin (CTX) and/or PTX. Studies with transducin α

identified an arginine at the approximate center of the molecules as the amino acid ADP-ribosylated by CTX (VanDop et al., 1984) and a cysteine at position −4 from the carboxyl-terminal end as the site of ADP-ribosylation by PTX (West et al., 1985). It is worth mentioning that not all G proteins have been either purified or cloned, and that not all the known G proteins, such as G_{o1} and G_{o2} or some of the G_i's, have unequivocally assigned functions, or that all functionally recognized G proteins, such as the inhibitory G_i of adenylyl cyclase or the stimulatory G_p of phospholipases, have been identified.

Functional Studies
Combined Use of Natural and Recombinant α Subunits Made in Bacteria to Define Their Functions

The abundance at which different α_s and α_i molecules are expressed varies from tissue to tissue, raising the question as to whether functional differences are associated with the structural differences, or whether the G_s's and the G_i's, respectively, should be thought of merely as isoforms. Although the final word on these questions is not yet in, we have during this last year developed the method(s) required to answer them. Thus, we have so far succeeded in purifying two types of G_i from human erythrocytes (hRBCs) and a third from bovine brain, from which we also purified the two forms of G_o. This allowed us to test for potential differential biological functions. We also expressed biologically active forms of the α subunits in bacteria, designated as recombinant α subunits, so that we could predict/confirm biological functions of various cloned and/or purified G protein α subunits.

We have not yet carried out all of the studies. However, we were able to determine first of all that both the natural puriifed and the recombinant forms of α_{i3}, α_{i1}, and α_{i2}, all stimulate K$^+$ channels (Mattera et al., 1989b; Yatani et al., 1988c). We failed to observe significant differences between types 1, 2, and 3 α_i molecules, even though the recombinant forms all had potencies between 30- and 50-fold lower than their natural counterparts (Yatani et al., 1988c). Thus, with respect to atrial muscarinic K$^+$ channels, G_{i1}, G_{i2}, G_{i3} must be considered iso-G proteins. Studies are currently in progress in atrial membrane patches in which the endogenous G_k has been uncoupled from receptors by treatment with PTX (Yatani et al., 1987c; Brown and Birnbaumer, 1988), to determine whether muscarinic and/or adenosine receptors exhibit selectivity for interaction with one or another of the G_i proteins.

Second, by testing the effect of recombinant α_s, we were able to determine that the stimulation of Ca^{2+} channels obtained with purified hRBC G_s (Imoto et al., 1988; Yatani et al., 1987b, 1988b) is indeed mediated by G_s as opposed to being due to a contaminant. Further, in collaboration with Michael Graziano and Al Gilman (University of Texas Medical School at Dallas) we tested the recombinant forms of three of the possible four splice variants of α_s for their Ca^{2+} channel stimulatory activity and found that they all do so with indistinguishable potency and efficacy (Mattera et al., 1989a). As was the case with recombinant α_i subunits, the recombinant α_s subunits also displayed a 20-fold reduced potency with respect to that of native human erythrocyte α_s. This applied not only to Ca^{2+} channel stimulation, but equally to adenylyl cyclase stimulation. It is very likely that bacteria fail to carry out a critical posttranslational

Gating of Ionic Channels as a Tool to Discover New Roles for G Proteins: Effects of G_o on Neuronal K^+ Channels

One of the properties of the "muscarinic" K^+ channels is that they are essentially silent in the absence of stimulation by an activated G protein (G_k). That is, in the absence of activated G protein their P_o is close to zero. The possibility existed that not only G_i proteins regulate K^+ channels but also the structurally closely related G_o. Since nerve tissue is rich in G_o, central nervous system neurons, specifically hippocampal pyramidal cells, were placed into culture and studied for the potential presence of both G_i- and G_o-gated K^+ channels. Although these studies are still in progress (VanDongen, T., L. Birnbaumer, and A. M. Brown, unpublished observations), the initial findings with highly purified bovine brain GTPγS-activated G_{o1} (G_{o1}^*) were of interest (VanDongen et al., 1988). They identified the existence of several novel G protein–gated, more precisely G_o-gated K^+, channels that are distinct from G_i-gated K^+ channels. Thus, application of purified bovine brain G_{o1}^* to the cytoplasmic aspect of inside-out membrane patches of cultured hippocampal pyramidal cells resulted in the appearance of three new types of single-channel K^+ currents consistent with the existence of the three nonrectifying types of K^+ channels having sizes of 13, 40, and 55 pS, respectively, plus an inwardly rectifying K^+ channel with a slope conductance of 40 pS. No such channel activities were observed with hRBC G_i^*-3 or hRBC α_i^*-3. G_{o2}^* or α_{o2}^* have not been tested in this system as yet. In contrast to earlier observations with G_o^* added to guinea pig atrial membrane patches, which showed only marginal effects of GTPγS-activated G_{o1} at 2 nM (Yatani et al., 1987a), the hippocampal K^+ channel is highly sensitive to G_o^* (Van Dongen et al., 1988), and significant activation was obtained at 1 pM and half-maximal effects were obtained at ~10 pM. The identity of the active G protein in the G_{o1} preparations used was confirmed with recombinant GTPγS-activated α_{o1}. The G_o-gated channels were stimulated in the absence of Ca^{2+} or ATP, in the presence or absence of AMP-P(NH)P, which was added routinely to inhibit ATP-sensitive 70-pS K^+ channels. Furthermore EGTA did not interfere with the actions of G_{o1} or *recombinant* α_{o1}. Thus, in hippocampal pyramidal cells of the rat, G_{o1} is a G_k, and the K^+ channels gated by it are several and differ from those present in atrial cells in various aspects including G protein specificity. These findings raise the question, which if any of the G proteins that gate K^+ channels regulate the other known PTX-sensitive effector systems.

βγ Dimers Inhibit K^+ Channel Gating by G Protein

Logothetis et al. (1987, 1988) and Kurachi et al. (1989a) reported that βγ dimers stimulate atrial K^+ channel activity. Their finding is not reproduced in our hands. On the contrary, when we add βγ dimers to inside-out membrane patches in which K^+ channels have been stimulated either by GTP only (baseline activity) or by carbachol plus GTP (agonist-stimulated activity), we consistently find inhibition of activity (Fig. 2). On their own, i.e., when added to silent patches in the absence of GTP, βγ dimers have no effect under our assay conditions (Fig. 1; Codina et al., 1987b; Kirsch et al., 1988; Okabe et al., 1990).

Concentration effect studies showed clearly that $\beta\gamma$ dimers are more potent in inhibiting agonist-independent than agonist-stimulated activity, and that this phenomenon applies not only to $\beta\gamma$ dimers suspended in Lubrol PX, such as those from human placenta, human erythrocytes, and bovine brain, but also to $\beta\gamma$ dimers presented to the patches in aqueous media, such as transducin $\beta\gamma$ (Fig. 3).

The fact that $\beta\gamma$ dimers inhibit GTP-dependent activity in the absence of agonist at lower concentrations, as compared with inhibition in the presence of agonist, was

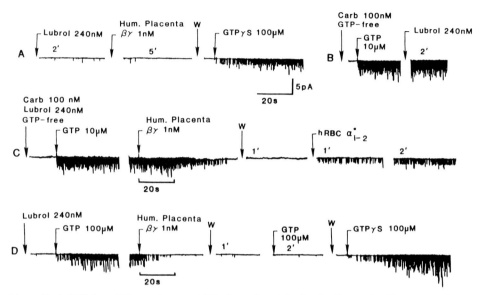

Figure 2. Inhibition of G protein–gated K$^+$ channel activity by $\beta\gamma$ dimers. Inside-out membrane patches from adult guinea pig atrial cells were exposed to the bathing solution (140 mM KCl, 2 mM MgCl$_2$, 5 mM EGTA, 10 mM HEPES-K, pH 7.4) containing the additives shown on the figure. The pipette solution was identical to the bathing solution and contained 100 nM carbachol when indicated. The composition of bathing solution was changed by a concentration-clamp method. Note that Lubrol PX and/or bovine serum albumin, used to maintain $\beta\gamma$ dimers in suspension, do not interfere with stimulation of activity by GTP (*D*) or GTP plus agonist (*B, C*), and that $\beta\gamma$ dimers have no effect on their own (*A*) but inhibit atrial K$^+$ channel stimulation by the membrane G_k (*C, D*). Inhibition is faster and is elicited with lower concentrations of $\beta\gamma$ dimers when K$^+$ channels are operating under baseline conditions (GTP only, experiment *D*: cessation of activity after 16 s) than when they are stimulated by agonist (carbachol plus GTP, experiment *C*: cessation of activity after 50 s). Numbers above records denote time elapsed in minutes between a solution change and the beginning of the record shown.

observed previously in purified components reconstituted into phospholipid vesicles (Birnbaumer, 1987), and serves to support our thesis that $\beta\gamma$ dimers act in intact membranes as suppressors of "noise" generated by agonist-unoccupied receptors.

The inhibitory effects of $\beta\gamma$ dimers obtained by us need to be contrasted to stimulatory effects obtained by Clapham, Neer, and their collaborators (Logothetis et al., 1987, 1988; Kim et al., 1989). We do not understand exactly the reasons for the discrepancy. We obtain inhibition also in the absence of Lubrol PX using $\beta\gamma$ dimers from transducin which are water soluble. The claim that $\beta\gamma$ dimers may be acting by stimulation of arachidonic acid formation (Kim et al., 1989), is suspect. In an adjacent report, Kurachi et al. (1989b) have shown that arachidonic acid and its metabolites require the presence of GTP for their action, and that they are blocked by GDPβS. Yet the stimulatory effects of $\beta\gamma$ dimers occur in the absence of GTP (Logothetis et al., 1987, 1988; Kim et al., 1989).

Current Views on How Signal Transduction by GTP and Receptors Comes about and Which Receptor Acts on Which G Protein to Regulate Which Effector System

Taken together the results discussed in the previous sections lead to several conclusions: (a) ionic channels are targets of direct regulation by G proteins as are adenylyl cyclase

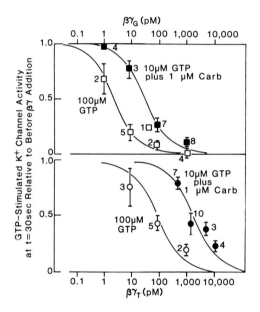

Figure 3. Effect of agonist (carbachol) on the dose-dependent inhibition by $\beta\gamma$ dimers of GTP-dependent K^+ channel activities in inside-out guinea pig atrial membrane patches by $\beta\gamma$ dimers. When 1 μM carbachol was present in the pipette, GTP was 10 μM; in the absence of carbachol GTP was 100 μM. $\beta\gamma_G$: data obtained with $\beta\gamma$ dimers derived from either human erythrocyte, human placenta, or bovine brain were pooled. $\beta\gamma_T$: data obtained with $\beta\gamma$ derived from bovine rod outer segments. We thank Dr. Tony Evans for the gift of human placental $\beta\gamma$ dimers, and Dr. J.-K. Ho, for their gift of bovine rod outer segment $\beta\gamma$ dimers. (Adapted from Okabe et al., 1990).

and the cGMP-specific phosphodiesterase; (b) α subunits and not $\beta\gamma$ dimers are the specificity determinants of signal transduction pathways; (c) several G proteins may have the same function, e.g., stimulation of K^+ channels by three G_i's; and (d) a single G protein may have more than one function, i.e., be multifunctional (e.g., stimulation of adenylyl cyclase and the DHP-sensitive Ca^{2+} channel by a single G_s).

Two independent sets of questions emerge from these findings. The first deals with the subunit dissociation reaction, and asks what type of advantage it confers onto the system by its existence. An answer to these questions can be found upon analyzing in detail the G protein regulatory cycle and the mechanism by which receptors promote G protein activation by GTP.

The second set of questions deals with crosstalk between signal transduction pathways, i.e., whether receptors act on more than one G protein, and if so which, whether G proteins interact with more than one receptor as well as with more than one effector, and if so, how frequent this is.

Role of Subunit Dissociation: Requirement for Catalytic Action of Receptors

In the second half of the 1970's it was demonstrated that receptors act catalytically rather than stoichiometrically to activate adenylyl cyclase (Tolkovsky and Levitzki,

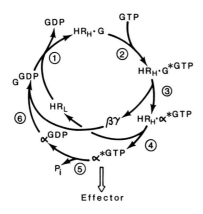

Figure 4. Integrated view of receptor-mediated catalytic activation of a G protein in the context of the dual subunit dissociation and GTPase cycles of G proteins. The role of receptors is to promote nucleotide exchange and to stabilize a GTP-dependent "activated" form of the G protein and the G protein undergoes a cyclical dissociation-reassociation reaction and oscillates between GDP, nucleotide-free, and GTP states. The cycles are driven energetically forward by the capacity of the G protein to hydrolyze GTP (Cassel and Selinger, 1978; Godchaux and Zimmerman, 1979; Brandt and Ross, 1986; Cerione et al., 1984), and kinetically by the dissociation of $\beta\gamma$ dimer from the activated receptor–G proteins complex. The receptor has high affinity for agonist (R_H) when associated with the nucleotide-free trimeric $\alpha\beta\gamma$ form of the G protein, and has low affinity for the agonist (R_L) when it is free. Further, the receptor has higher affinity for the trimeric $\alpha\beta\gamma$ form of G than the G-GDP, thus accounting for the finding that GDP and GDP analogues promote the R_H to R_L transition. The $\beta\gamma$ dimers are required for the interaction of α with R, and after formation of G-GTP, no G* forms unless it is "aided" by receptor. For this receptor has to have an even higher affinity for the G*-GTP state than the nucleotide-free state of G. As a consequence, receptor dissociation is absolutely dependent on reaction 3 (subunit dissociation). (Thermodynamic reasons do not allow R both to stabilize the G* state and to dissociate from it.) Reaction 4 states further that the $R_H\alpha$*GTP loses its ability to stay associated and decomposes further into free activated α*GTP plus free receptor, thus accounting for the fact that under "working" conditions (saturation by both GTP and hormone, and hence sustained regulation of effector) only a small proportion of receptors are found in their high affinity, G protein–associated state. It follows that the G protein cycle is driven forward not only by the GTPase but also, and obligatorily so, by the subunit dissociation reaction. The scheme accounts for the experimental findings (a) that receptors act catalytically and therefore need to dissociate from the G protein at one time or another (Tolkovsky and Levitzki (1978), and (b) that receptors accelerate the transition from inactive G-GTPγS to active G*-GTPγS transition and therefore must have higher intrinsic affinity for the activated than the inactive state (Birnbaumer et al., 1980; Iyengar et al., 1980; Iyengar and Birnbaumer, 1982).

1978) and in 1980 it was found that receptors, in addition to promoting GDP/GTP exchange (Cassel and Selinger, 1978; Cassel et al., 1979), also participate in promoting the actual activation reaction of adenylyl cyclase by GTP and its analogues (reviewed in Birnbaumer et al., 1985). These two findings, catalytic action and stabilization of the nucleotide-activated form of adenylyl cyclase, now known to be in fact G_s, are thermodynamically impossible, unless the stabilized form of the G protein undergoes some type of additional spontaneous change that releases it from microscopic reversibility constraints, in analogy to what happens in all enzymatically catalyzed reactions in which the product is chemically different from the starting substrate. Dissociation of $\beta\gamma$ from the activated and stable receptor–G protein complex is such a change. We therefore propose that the role of the dissociation reaction is a requirement without which receptors cannot act catalytically. Both the catalytic nature of receptor-mediated activation of G_s and the fact that the receptor affects not only GDP release but also the rate at which inactive guanine nucleotide–occupied G protein isomerizes from an inactive to an active state have been confirmed in reconstitution experiments with purified receptor and purified G_s (Asano et al., 1984; Brandt and Ross, 1986). Since receptors do not interact with the α subunits except in the context of $\beta\gamma$ (Kanaho et al., 1984; Florio and Sternweis, 1985, 1989; Watkins et al., 1985; Tota et al., 1987), reassociation of the GTPase-deactivated α with $\beta\gamma$ is essential for restimulation by receptors. This then leads to a description of the G protein regulatory cycle under the influence of receptor as depicted in Fig. 4.

Taken together, it is thus possible to ascribe three generic functions to $\beta\gamma$ dimers (Table III): (a) activation of α subunit by receptor, for without $\beta\gamma$ receptors do not interact with α; (b) amplification of the receptor signal, for without dissociation receptors cannot act catalytically; and (c) noise reduction, for unoccupied receptors are not silent.

TABLE III
Intramembrane Roles of G Protein $\beta\gamma$ Dimers

Reaction	Product	Role
Reassociation with α^{GDP}	G^{GDP}	Activation by R
Dissociation from HR.G*^{GTP}	HR.α*^{GTP}	Signal amplification
Reassociation with R.α*^{GTP}	R.G^{GTP}	Noise reduction

Specificities in Receptor–G Protein–Effector Interaction

Ever since the discoveries in the late 1960's that up to five different hormone receptors can activate in a single adenylyl cyclase system in an isolated membrane (Birnbaumer and Rodbell, 1969) and, in the early and mid 1970's, that receptors can be transferred from one cell to another (Citri and Schramm, 1975), and that there are no species and/or tissue specificity restrictions as to the source of G_s for reconstituion of a hormonally stimulatable adenylyl cyclase system in cyc- membranes (Ross et al., 1978; Kaslow et al., 1979), it has been clear that single G proteins are designed to interact with classes of receptors as opposed to single receptor subtypes . Experimental results mentioned above indicate that one G protein can interact with more than one effector and that several G proteins may regulate a single effector. Finally, Ashkenazi et al. (1987) showed that single receptors may effect more than one G protein. The important notion that emerges from these considerations is that the wiring diagram describing signal transduction by G proteins needs to be determined individually and separately for each cell or tissue of interest. This includes the determination not only of which receptors are present but also of which G proteins and which effector functions are expressed.

Acknowledgments

Supported in part by United States Public Health Service research grants DK-19318, HD-09581, HL-31164, and HL-37044 to Lutz Birnbaumer, HL-39262 to Dr. Arthur M. Brown (Department of Molecular Physiology and Biophysics), a Welch Foundation grant (Q1075), and the Baylor College of Medicine Diabetes and Endocrinology Research Center grant DK-27685.

References

Asano, T., S. E. Pedersen, C. W. Scott, and E. M. Ross. 1984. Reconstitution of catecholamine-stimulated binding of guanosine 5'-o-(3-thiotriphosphate) to the stimulatory GTP-binding protein of adenylate cyclase. *Biochemistry.* 23:5460–5467.

Ashkenazi, A., J. W. Winslow, E. G. Peralta, G. L. Peterson, M. I. Schimerlik, D. J. Capon, and J. Ramachandran. 1987. An M2 muscarinic receptor subtype coupled to both adenylyl cyclase and phosphoinositide turnover. *Science.* 238:672–675.

Birnbaumer, L. 1987. Which G protein subunits are the active mediators in signal transduction. *Trends in Pharmacological Science.* 8:209–211.

Birnbaumer, L. 1990. G proteins in signal transduction. *Annual Reviews of Pharmacology and Toxicology.* In press.

Birnbaumer, L., J. Codina, A. Yatani, R. Mattera, R. Graf, J. Olate, A. P. N. Themmen, C.-F. Liao, J. Sanford, K. Okabe, Y. Imoto, Z. Zhou, J. Abramowitz, W. S. Suki, H. E. Hamm, R. Iyengar, M. Birnbaumer, and A. M. Brown. 1989. Molecular basis of regulation of ionic channels by G proteins. *Recent Progress in Hormone Research.* 45:120–208.

Birnbaumer, L., J. D. Hildebrandt, J. Codina, R. Mattera, R. A. Cerione, T. Sunyer, F. J. Rojas, M. G. Caron, R. J. Lefkowitz, and R. Iyengar. 1985. Structural basis of adenylate cyclase stimulation and inhibition by distinct guanine nucleotide regulatory proteins. *In* Molecular Mechanisms of Signal Transduction. P. Cohen, and M. D. Houslay, editors. Elsevier/North Holland Biomedical Press, Amsterdam. 131–182.

Birnbaumer, L., and M. Rodbell. 1969. Adenyl cyclase in fat cells. II. Hormone receptors. *Journal of Biological Chemistry.* 244:3477–3482.

Birnbaumer, L., T. L. Swartz, J. Abramowitz, P. W. Mintz, and R. Iyengar. 1980. Transient and steady state kinetics of the interaction of nucleotides with the adenylyl cyclase system from rat liver plasma membranes: interpretation in terms of a simple two-state model. *Science.* 255:3542–3551.

Breitweiser, G. E., and G. Szabo. 1985. Uncoupling of cardiac muscarinic and beta-adrenergic receptors from ion channels by a guanine nucleotide analogue. *Nature.* 317:538–540.

Brandt, D. R., and E. M. Ross. 1986. Catecholamine-stimulated GTPase cycle. Multiple sites of regulation by beta-adrenergic receptor and Mg^{2+} studied in reconstituted receptor-G_s vesicles. *Journal of Biological Chemistry.* 261:1656–1664.

Brown, A. M., and L. Birnbaumer. 1988. Direct G protein gating of ion channels. *American Journal of Physiology.* 23:H401–H410.

Cassel, D., F. Eckstein, M. Lowe, and Z. Selinger. 1979. Determination of the turn-off reaction for the hormone-activated adenylate cyclase. *Journal of Biological Chemistry.* 254:9835–9838.

Cassel, D., and Z. Selinger. 1978. Mechanism of adenylate cyclase activation through the beta-adrenergic receptor: catecholamine-induced displacement of bound GDP by GTP. *Proceedings of the National Academy of Sciences.* 75:4155–4159.

Cerione, R. A., J. Codina, J. L. Benovic, R. J. Lefkowitz, L. Birnbaumer, and M. G. Caron. 1984. The mammalian beta$_2$-adrenergic receptor: reconstitution of the pure receptor with the pure stimulatory nucleotide binding protein (N_s) of the adenylate cyclase system. *Biochemistry.* 23:4519–4525.

Citri, Y., and M. Schramm. 1975. Resolution, reconstitution and kinetics of the primary action of hormone receptor. *Nature.* 287:297–300.

Codina, J., D. Grenet, A. Yatani, L. Birnbaumer, and A. M. Brown. 1987a. Hormonal regulation of pituitary GH_3 cell K^+ channels by G_k is mediated by its alpha subunit. *FEBS Letters.* 216:104–106.

Codina, J., A. Yatani, D. Grenet, A. M. Brown, and L. Birnbaumer. 1987b. The alpha subunit of the GTP binding protein g_k opens atrial potassium channels. *Science.* 236:442–445.

Codina, J., J. Olate, J. Abramowitz, R. Mattera, R. G. Cook, and L. Birnbaumer. 1988. Alpha$_i$-3 cDNA encodes the alpha subunit of G_k, the stimulatory G protein of receptor-regulated K^+ channels. *Journal of Biological Chemistry.* 263:6746–6740.

Dolphin, A. C., R. H. Scott, and J. F. Wooton. 1988. Photorelease of GTPγS inhibits the low threshold channel current in cultured rat dorsal root ganglion (DRG) neurons. *Pflügers Archiv.* 411:19P.

Ewald, D. A., I.-H. Pang, P. C. Sternweis, and R. J. Miller. 1989. Differential G protein-mediated coupling of neurotransmitter receptors to Ca^{2+} channels in rat dorsal root ganglion neurons in vitro. *Neuron.* 2:1185–1193.

Ewald, D. A., P. C. Sternweis, and R. J. Miller. 1988. Guanine nucleotide-binding protein G_o-induced coupling of neuropeptide Y receptors to Ca^{2+} channels in sensory neurons. *Proceedings of the National Academy of Sciences.* 85:3633–3637.

Florio, V. A., and P. C. Sternweis. 1985. Reconstitution of resolved muscarinic cholinergic receptors with purified GTP-binding proteins. *Journal of Biological Chemistry.* 260:3477–3483.

Florio, V. A., and P. C. Sternweis. 1989. Mechanism of muscarinic receptor action on G_o in reconstituted phospholipid vesicles. *Journal of Biological Chemistry.* 264:3909–3915.

Godchaux, W., III, and W. F. Zimmerman. 1979. Membrane-dependent guanine nucleotide binding and GTPase activities of soluble protein from bovine rod cell outer segments. *Journal of Biological Chemistry.* 254:7874–7884.

Hescheler, J., W. Rosenthal, K.-D. Hinsch, M. Wulfern, W. Trautwein, and G. Schultz. 1988. Angiotensin II-induced stimulation of voltage-dependent Ca^{2+} currents in an adrenal cortical cell line. *EMBO J.* 7:619–624.

Hescheler, J., W. Rosenthal, W. Trautwein, and G. Schultz. 1987. The GTP-binding protein, N_o, regulates neuronal calcium channels. *Nature.* 325:445–447.

Holz, G. G., S. G. Rane, and K. Dunlap. 1986. GTP-binding proteins mediate transmitter inhibition of voltage-dependent calcium channels. *Nature.* 319:670–672.

Imoto, Y., A. Yatani, J. P. Reeves, J. Codina, L. Birnbaumer, and A. M. Brown. 1988. α subunit of G_s directly activates cardiac calcium channels in lipid bilayers. *American Journal of Physiology.* 255:H722–H728.

Iyengar, R., J. Abramowitz, M. Riser, and L. Birnbaumer. 1980. Hormone receptor-mediated stimulation of the rat liver plasma membrane adenylyl cyclase system: nucleotide effects and analysis in terms of a two-state model for the basic receptor-affected enzyme. *Journal of Biological Chemistry*. 255:3558–3564.

Iyengar, R., and L. Birnbaumer. 1982. Hormone receptors modulate the regulatory component of adenylyl cyclases by reducing its requirement for Mg ion and increasing its extent of activation by guanine nucleotides. *Proceedings of the National Academy of Sciences*. 79:5179–5183.

Kameyama, M., J. Hescheler, F. Hofmann, and W. Trautwein. 1986. Modulation of Ca current during the phosphorylation cycle in the guinea pig heart. *Pflügers Archiv*. 407:121–128.

Kanaho, Y., S.-C. Tsai, R. Adamik, E. L. Hewlett, J. Moss, and M. Vaughan. 1984. Rhodopsin-enhanced GTPase activity of the inhibitory GTP-binding protein of adenylate cyclase. *Journal of Biological Chemistry*. 159:7378–7381.

Kaslow, H. R., Z. Farfel, G. L. Johnson, and H. R. Bourne. 1979. Adenylate cyclase assembled in vitro: cholera toxin substrates determine different patterns of regulation by isoproterenol and guanosine 5'-triphosphate. *Molecular Pharmacology*. 15:472–483.

Kim, D., D. L. Lewis, L. Graziadei, E. J. Neer, D. Bar-Sagi, and D. E. Clapham. 1989. G protein $\beta\gamma$-subunits activate the cardiac muscarinic K^+ channel via phospholipase A_2. *Nature*. 337:557–560.

Kirsch, G., A. Yatani, J. Codina, L. Birnbaumer, and A. M. Brown. 1988. The alpha subunit of G_k activates atrial K^+ channels of chick, rat and guinea pig. *American Journal of Physiology*. 254:(*Heart Circulation Physiology*. 23):H1200–H1205.

Kurachi, Y., H. Ito, T. Sugimoto, T. Katada, and M. Ui. 1989a. Activation of atrial muscarinic K^+ channels by low concentrations of $\beta\gamma$ subunits of G rat brain protein. *Pflügers Archiv*. 413:325–327.

Kurachi, Y., H. Ito, T. Sugimoto, T. Shimizu, I. Miki, and M. Ui. 1989b. Arachidonic acid metabolites as intracellular modulators of the G protein-gated cardiac K^+ channel. *Nature*. 337:555–557.

Lewis, D. L., F. F. Weight, and A. Luini. 1986. A guanine nucleotide-binding protein mediates the inhibition of voltage-dependent calcium current by somatostatin in a pituitary cell line. *Proceedings of the National Academy of Sciences*. 83:9035–9039.

Light, D. B., D. Ausiello, and B. A. Stanton. 1989. Guanine nucleotide binding protein, α_i^*-3, directly activates a cation channel in rat inner medullary collecting duct cell. *Journal of Clinical Investigation*. 84:354–356.

Lochrie, M. A., and M. I. Simon. 1988. G protein multiplicity in eukaryotic signal transduction systems. *Biochemistry*. 17:4957–4965.

Logothetis, D. E., D. Kim, J. K. Northup, E. J. Neer, and D. E. Clapham. 1988. Specificity of action of guanine nucleotide-binding regulatory protein subuntis on the cardiac muscarinic K^+ channel. *Proceedings of the National Academy of Sciences*. 85:5814–5818.

Logothetis, D. E., Y. Kurachi, J. Galper, E. J. Neer, and D. E. Clapham. 1987. The $\beta\gamma$ subunits of GTP-binding proteins activate the muscarinic K^+ channel in heart. *Nature*. 325:321–326.

Mattera, R., M. P. Graziano, A. Yatani, Z. Zhou, R. Graf, J. Codina, L. Birnbaumer, A. G. Gilman, and A. M. Brown. 1989a. Individual splice variants of the α subunit of the G protein G_s activate both adenylyl cyclase and Ca^{2+} channels. *Science*. 243:804–807.

Mattera, R., A. Yatani, G. E. Kirsch, R. Graf, J. Olate, J. Codina, A. M. Brown, and L. Birnbaumer. 1989b. Recombinant α_i-3 subunit of G protein activates G_k-gated K^+ channels. *Journal of Biological Chemistry*. 264:465–471.

Nargeot, J., J. M. Nerbonne, J. Engels, and H. A. Lester. 1983. Time course of the increase in the myocardial slow inward current after a photochemically generated concentration jump of intracellular cAMP. *Proceedings of the National Academy of Sciences*. 80:2395–2399.

Okabe, K., A. Yatani, T. Evans, K.-J. Ho, J. Codina, L. Birnbaumer, A. M. Brown. 1990. $\beta\gamma$ dimers of G proteins: Inhibition of receptor-mediated activation of atrial K^+ channels and lack of specificity for interaction with α subunits. *Journal of Biological Chemistry*. In press.

Parent, L., and R. Coronado. 1989. Reconstitution of the ATP-sensitive potassium channel of skeletal muscle. Activation of a G-protein dependent process. *Journal of General Physiology*. 94:445–463.

Pfaffinger, P. J., J. M. Martin, D. D. Hunter, N. M. Nathanson, and B. Hille. 1985. GTP-binding proteins couple cardiac muscarinic receptors to a K channel. *Nature*. 317:536–538.

Ribalet, B., S. Ciani, and G. T. Eddlestone. 1989. Modulation of ATP-sensitive K channels in RINm5F cells by phosphorylation and G proteins. *Biophysical Journal*. 55:587. (Abstr.)

Rosenthal, W., J. Hescheler, K.-D. Hinsch, K. Spicher, W. Trautwein, and G. Schultz. 1988. Cyclic AMP-independent, dual regulation of voltage-dependent Ca^{2+} currents by LHRH and somatostatin in a pituitary cell line. *EMBO J*. 7:1627–1633.

Ross, E. M., A. C. Howlett, K. M. Ferguson, and A. G. Gilman. 1978. Reconstitution of hormone-sensitive adenylate cyclase activity with resolved components of the enzyme. *Journal of Biological Chemistry*. 253:6401–6412.

Schubert, B., A. M. J. VanDongen, G. E. Kirsch, and A. M. Brown. 1989. Modulation of cardiac Na channels by beta-adrenoreceptors and the G protein G_s. *Biophysical Journal*. 55:229. (Abstr.)

Scott, R. H., and A. C. Dolphin. 1986. Regulation of calcium currents by GTP analogue: potentiation of (-)-baclofen-mediated inhibition. *Neuroscience Letters*. 69:59–64.

Soejima, M., and A. Noma. 1984. Mode of regulation of the ACh-sensitive K-channel by the muscarinic receptor in rabbit atrial cells. *Pflügers Archiv*. 400:424–431.

Tolkovsky, A. M., and A. Levitzki. 1978. Mode of coupling between the β-adrenergic receptor and adenylate cyclase in turkey erythrocytes. *Biochemistry*. 17:3795–3810.

Toro, L., J. Ramos-Franco, and E. Stefani. 1990. GTP-dependent regulation of myometrial K_{ca} channels incorporated into lipid bilayers. *Journal of General Physiology*. 96: in press.

Tota, M. R., K. R. Kahler, and M. I. Schimerlik. 1987. Reconstitution of the purified porcine atrial muscarinic acetylcholine receptor with purified porcine atrial inhibitory guanine nucleotide binding protein. *Biochemistry*. 26:8175–8182.

Van Dop, C., M. Tsubokawa, H. R. Bourne, and J. Ramachandran. 1984. Amino acid sequence of retinal transducin at the site ADP-ribosylated by cholera toxin. *Journal of Biological Chemistry*. 259:696–699.

VanDongen, A., J. Codina, J. Olate, R. Mattera, R. Joho, L. Birnbaumer, and A. M. Brown. 1988. Newly identified brain potassium channels gated by the guanine nucleotide binding (G) protein G_o. *Science*. 242:1433–1437.

Watkins, P. A., D. L. Burns, Y. Kanaho, T.-Y. Liu, E. L. Hewlett, and J. Moss. 1985. ADP-ribosylation of transducin by pertussis toxin. *Journal of Biological Chemistry.* 260:13478–13482.

West, R. E., Jr., J. Moss, M. Vaughan, T. Liu, and T.-Y. Liu. 1985. Pertussis toxin–catalyzed ADP-ribosylation of transducin. Cystein 347 is the ADP-ribose acceptor site. *Journal of Biological Chemistry.* 260:14428–14430.

Yatani, A., J. Codina, A. M. Brown, and L. Birnbaumer. 1987a. Direct activation of mammalian atrial muscarinic potassium channels by GTP regulatory protein G_k. *Science.* 235:207–211.

Yatani, A., J. Codina, Y. Imoto, J. P. Reeves, L. Birnbaumer, and A. M. Brown. 1987b. A G protein directly regulates mammalian cardiac calcium channels. *Science.* 238:1288–1292.

Yatani, A., J. Codina, R. D. Sekura, L. Birnbaumer, and A. M. Brown. 1987c. Reconstitution of somatostatin and muscarinic receptor mediated stimulation of K^+ channels by isolated G_k protein in clonal rat anterior pituitary cell membranes. *Molecular Endocrinology.* 1:283–289.

Yatani, A., H. Hamm, J. Codina, M. R. Mazzoni, L. Birnbaumer, and A. M. Brown. 1988a. A monoclonal antibody to the α subunit of G_k blocks muscarinic activation of atrial K^+ channels. *Science.* 241:828–831.

Yatani, A., Y. Imoto, J. Codina, S. L. Hamilton, A. M. Brown, and L. Birnbaumer. 1988b. The stimulatory G protein of adenylyl cyclase, G_s, directly stimulates dihydropyridine-sensitive skeletal muscle Ca^{2+} channels. Evidence for direct regulation independent of phosphorylation by cAMP-dependent protein kinase. *Journal of Biological Chemistry.* 263:9887–9895.

Yatani, A., R. Mattera, J. Codina, R. Graf, K. Okabe, E. Padrell, R. Iyengar, A. M. Brown, and L. Birnbaumer. 1988c. The G protein-gated atrial K^+ channel is stimulated by three distinct $G_i\alpha$-subunits. *Nature.* 336:680–682.

Chapter 16

Antibodies against Synthetic Peptides as Probes of G Protein Structure and Function

Allen M. Spiegel, William F. Simonds, Teresa L. Z. Jones, Paul K. Goldsmith, and Cecilia G. Unson

Molecular Pathophysiology Branch, National Institute of Diabetes, Digestive, and Kidney Diseases, National Institutes of Health, Bethesda, Maryland 20892; and Department of Biochemistry, Rockefeller University, New York, New York

Introduction

Five different antisera have been raised against synthetic decapeptides corresponding to the carboxy terminus of G protein α subunits. The specificity of these antisera has been defined by peptide ELISA, immunoblots, and immunoprecipitation of recombinant G proteins expressed in *Escherichia coli* and immunoblots of purified G proteins. Antisera raised against the carboxy-terminal decapeptides of α_s amd α_z were absolutely specific for the corresponding G proteins. An antiserum raised against the carboxy-terminal decapeptide of α_t showed strong reactivity against α_{i1} and α_{i2}, but only weak reactivity with α_{i3}. Antisera raised against carboxy-terminal decapeptides of α_{i3} and α_o showed strong reactivity against the cognate G protein, but also showed substantial reciprocal cross-reactivity. Antisera raised against α subunit carboxy-terminal decapeptides proved capable of immunoprecipitating their cognate G proteins, as well as recognizing these proteins in native cell membranes. Thus the α_s-specific antiserum could block agonist-stimulated adenylyl cyclase activity in native membranes, and could also immunoprecipitate an activated α_s-adenylyl cyclase complex. The α_{i2}-, but not α_{i3}- and α_z-specific antiserum could block agonist-mediated inhibition of adenylyl cyclase in human platelet membranes. These results indicate that the carboxy-terminal decapeptide is involved in G protein receptor, but not effector, coupling. These antisera also proved useful in immunoprecipitation of endogenous and transfected α subunits in COS (monkey kidney) cells. Using this approach we were able to show that both α_s and α_i are membrane associated, but only the latter is myristylated. A mutant α_{i1} (second residue gly changed to ala) fails to incorporate myristate and is localized in the soluble fraction. Myristylation is, thus, essential for membrane attachment of α_i. These studies illustrate the utility of antisera raised against α subunit carboxy-terminal decapeptides in studies of G protein structure and function.

General Features of G Protein Structure and Function

G proteins involved in signal transduction are members of a guanine nucleotide–binding protein superfamily that includes cytoskeletal proteins such as tubulin, soluble proteins (initiation and elongation factors involved in protein synthesis), and low molecular weight GTP-binding proteins such as the *ras* p21 protooncogenes and *ras*-related proteins (Gilman, 1987; Iyengar and Birnbaumer, 1987; Spiegel, 1987). Members of the G protein subset of the GTP-binding protein superfamily share certain general features in common with other members of the GTP-binding protein superfamily: (*a*) all GTP-binding proteins bind guanine nucleotides with high affinity and specificity, and possess intrinsic GTPase activity that modulates interactions between the GTP-binding protein and other elements; (*b*) GTP-binding proteins serve as substrates for ADP-ribosylation by bacterial toxins; this covalent modification disrupts normal function.

G proteins share other features that distinguish them from other GTP-binding proteins. These features include: (*a*) association with the cytoplasmic surface of the plasma membrane (*ras* p21 and some other low molecular weight GTP-binding proteins are also associated with the cytoplasmic membrane surface); (*b*) they function as receptor-effector couplers; and (*c*) their heterotrimeric structure. G proteins contain α, β, and γ subunits, each distinct gene products. The latter two subunits are tightly, but noncovalently linked in a $\beta\gamma$ complex. α subunits bind guanine nucleotide, serve as toxin substrates, confer specificity in receptor-effector coupling, and directly modulate effector activity. Upon activation by GTP, α subunits are thought to dissociate from the $\beta\gamma$ complex. The $\beta\gamma$ complex is required for G protein-receptor interaction, can inhibit G protein activation by blocking α subunit dissociation, and may in some cases directly regulate effector activity.

Specific Features of G Protein Structure and Function

Molecular cloning provides evidence for a minimum of nine distinct α subunit genes, including G_s, G_o, G_{i1}, G_{i2}, G_{i3} G_{t1}, G_{t2} (Gilman, 1987; Iyengar and Birnbaumer, 1987; Spiegel, 1987), $G_{x(z)}$ (Fong et al., 1988; Lochrie and Simon, 1988; Matsuoka et al., 1988), and G_{olf} (Jones and Reed, 1989). Further diversity is created by alternative splicing leading to the expression of four forms of G_s (Bray et al., 1986). At least two distinct genes each exist for both β and γ subunits. The expression of certain α subunits is highly restricted, e.g., G_{t1} and G_{t2} are found only in photoreceptor rod and cone cells, respectively, and G_{olf} is only in olfactory neurons, whereas others such as G_s are expressed ubiquitously.

The specificity of G protein interactions with receptors and effectors has been defined in very few cases. Studies involving reconstitution of purified receptors and G proteins in phospholipid vesicles showed that β-adrenergic receptors couple, in decreasing order of efficiency, to $G_s > G_i \gg G_t$, and that for rhodopsin, the selectivity of coupling is $G_t = G_i \gg G_s$ (Cerione et al., 1985). Similar studies involving G protein-effector interaction indicated that only G_s can activate adenylyl cyclase (G_s also appears to stimulate another effector, a Ca^{++} channel), and that G_t uniquely activates retinal cGMP phosphodiesterase (see Roof et al., 1985 for example). The endogenous G proteins coupled to most other receptors and effectors, however, remain to be identified. Many G protein-coupled receptors (D_2-dopaminergic for example; Senogles et al., 1987), and a variety of effectors, including adenylyl cyclase (inhibition), certain Ca^{++} (inhibition) and K^+ (stimulation) channels, and phospholipase C (stimulation), in cell types such as neutrophils (e.g., by f-Met-Leu-Phe receptor) are regulated by one or more pertussis toxin–sensitive G proteins (Gilman, 1987; Iyengar and Birnbaumer, 1987; Spiegel, 1987). Because not only both forms of G_t, but also G_{i1}, G_{i2}, G_{i3}, and G_o are pertussis toxin–sensitive G proteins, demonstration of an effect of pertussis toxin on receptor or effector regulation does not uniquely identify the relevant endogenous G protein. In most cells, phospholipase C is regulated by a pertussis toxin–insensitive G protein. $G_{z(x)}$ may play this role, but definitive evidence is lacking.

Peptide Antibodies as Probes of G Protein Structure and Function
Generation and Characterization of Antisera

We have generated numerous antisera against synthetic peptides corresponding to the predicted amino acid sequence of G protein subunits. Such antisera have proved very useful in identifying the putative proteins encoded by cloned cDNAs (Goldsmith et al., 1987, 1988b), in discriminating between closely related G proteins (Goldsmith et al., 1988a), and in quantitating and localizing G proteins. For example, three antisera, LD, LE, and SQ were raised against an internal decapeptide (residues 159–168) of α subunits of G_{i1}, G_{i2}, and G_{i3}, respectively. These proved capable of discriminating between α subunits of 41 and 40 kD purified from brain, and of 41 kD purified from

HL-60 cells, and served to identify these as the products of G_{i1}, G_{i2}, and G_{i3}, respectively (Goldsmith et al., 1988a, b).

One approach theoretically capable of identifying endogenous G proteins coupling to particular receptors and effectors involves the use of specific antibodies that can bind to native G proteins and affect their function. Antibodies directed against the carboxy terminus of G_α subunits hold particular promise in this regard. Several lines of evidence point to the importance of the extreme carboxy terminus in G protein–receptor coupling. These include the locus of the receptor-uncoupling mutation (*unc*) in S49 mouse lymphoma cells (sixth residue from carboxy terminus; Sullivan et al., 1987), and the ability of pertussis toxin–catalyzed ADP-ribosylation of a cysteine residue (fourth from carboxy terminus) to uncouple G proteins from receptors (West et al., 1985).

To study G protein coupling with this approach, antibodies with well-defined specificity are essential. We had previously raised antisera AS/6 and 7, and GO/1 (referred to henceforth as AS and GO, see Table I) against synthetic decapeptides corresponding to the carboxy termini of α subunits of G_t and G_o, respectively. We

TABLE I
Carboxy-Terminal Decapeptides and Corresponding Antisera

Peptide	Sequence*	Antiserum[‡]	$G\alpha$ subunit[§]
RM	RMHLRQYELL	RM	G_s
QN	QNNLKYIGLC	QN	$G_{x(z)}$
GO	ANNLRGCGLY	GO	G_o
EC	KNNLKECGLY	EC	G_{i3}
KE	KENLKDCGLF	AS	G_{t1}, G_{t2}
KN	KNNLKDCGLF	—	G_{i1}, G_{i2}

*Single-letter amino acid code.
[‡]Antiserum raised by immunization with designated peptide.
[§]Designated peptide represents carboxy terminus of corresponding G_α subunit (Lochrie and Simon, 1988).

characterized the specificity of these antisera by performing immunoblots of several distinct purified G proteins. With this approach, we found that AS reacts, not only with G_t purified from rod outer segments (presumably G_{t1}), but also with α_{41} and α_{40} proteins (presumably G_{i1} and G_{i2}, respectively) purified from brain. We also showed that AS does not react with purified brain α_{39} (G_o), but does react, albeit weakly, with an α_{41} protein purified from HL-60 cells and tentatively identified as G_{i3} (Goldsmith et al., 1987 and 1988a). GO antiserum was shown to react with brain α_{39} (G_o), but not with brain α_{41} (G_{i1}) or α_{40} (G_{i2}); reactivity against G_{i3} had not been tested (Goldsmith et al., 1988b).

To generate antisera capable of recognizing each of the known G_α subunits, we synthesized additional decapeptides corresponding to the carboxy termini of G_s, $G_{x(z)}$, and G_{i3}, and used these to immunize rabbits (Table I). We initially assessed the reactivity and specificity of these antibodies against the immunogenic and related peptides by ELISA (Fig. 1). The results indicated that antisera RM and QN are highly specific for their cognate peptides, whereas AS, EC, and GO show substantial cross-reactivity for each other's immunogenic peptide. This presumptively reflects the

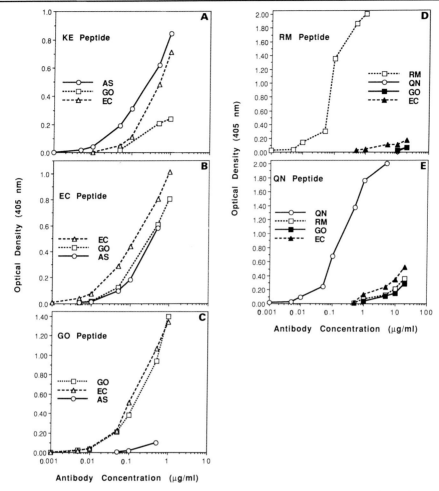

Figure 1. Characterization of specificity of G_α–carboxy-terminal decapeptide antisera by ELISA. ELISA was performed with plates coated with the indicated peptides (see Table I), and then incubated with increasing concentrations of the indicated affinity-purified antibodies. For plates coated with the KE, EC, and GO peptides, affinity-purified RM and QN antibodies (up to 10 μg/ml) gave no detectable reactivity. For plates coated with RM and QN peptides, affinity-purified AS antibodies (up to 20 μg/ml) gave no detectable reactivity. The values are the means of quadruplicate determinations.

relatively unique sequences of the RM and QN peptides vs. the more homologous sequences of the AS, EC, and GO peptides (Table I). We next tested the reactivity of the antisera with proteins encompassing the immunogenic peptide sequences by performing immunoblots and immunoprecipitation of bacterial lysates containing unique, defined G_α subunits expressed by recombinant DNA techniques. On immunoblot, RM and QN antisera are absolutely specific for G_s and G_x, respectively. G_s and G_x, moreover, are not reactive with GO, EC, or AS antisera. AS antiserum reacts strongly and equivalently with G_{i1} and G_{i2}, shows weak reactivity against G_{i3}, and no reactivity against G_o. GO and EC antisera react best with G_o and G_{i3}, respectively, but they display very substantial reciprocal cross-reactivity. The somewhat unexpected pattern of reactivity of AS, GO, and EC reflects the importance of the carboxy-terminal residue, phenylalanine vs. tyrosine, in determining antigenicity (Table I). The identical pattern of reactivity was observed in immunoprecipitation experiments which also indicated that the antisera can recognize native G proteins.

Peptide Antibodies as Probes of G Protein Receptor-Effector Coupling

Using purified proteins reconstituted into phospholipid vesicles, we found that affinity-purified AS antibodies specifically block rhodopsin-G_t (transducin) interaction, but do not block interaction between activated G_t and its effector, cGMP phosphodiesterase (Cerione et al., 1988). This approach was then extended to G proteins in situ, i.e., in native membranes (Simonds et al., 1989c). RM antibodies were shown on immunoblots to recognize the multiple forms of G_s-α derived from alternative splicing. Affinity-purified RM antibodies effectively block receptor-mediated adenylyl cyclase stimulation by the β-adrenergic agonist, isoproterenol, in S49 mouse lymphoma cell membranes. Fluoride stimulation (which bypasses the receptor to activate G_s) is only partially blocked by RM antibodies at much higher concentrations than those required for inhibition of isoproterenol-stimulated activity (Fig. 2). Antibody inhibition of

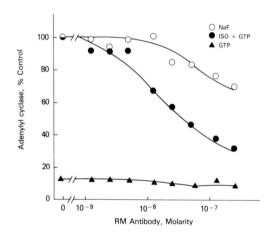

Figure 2. RM antibody inhibition of G_s activation in S49 mouse lymphoma cell membranes. Membranes were incubated with increasing concentrations (as shown) of RM affinity-purified antibody for 2 h at 4°C, after which aliquots were assayed for adenylyl cyclase activity with 10 mM NaF, or with 100 μM GTP with or without 500 μM l-isoproterenol (as shown). Values (in picomoles cAMP per minute per milligram) are the mean of triplicate determinations and are expressed as percent of control (84% for NaF and 60% for isoproterenol). (From Simonds et al., 1989c.)

adenylyl cyclase stimulation could be completely prevented by the cognate peptide. When membranes are preactivated with fluoride (not shown) or with GTP-γ-S (Fig. 3, lane D), RM antibodies specifically immunoprecipitate an active G_s–adenylyl cyclase complex.

In human platelet membranes, we identified G_s, G_{i2}, G_{i3}, and $G_{x(z)}$, but neither G_{i1} nor G_0, by immunoblot. As for S49 lymphoma cell membranes, RM antibodies, by binding to α_s, could block stimulation of adenylyl cyclase by agonist (prostaglandin E_1 [PGE_1] in platelet membranes). AS, but not EC or QN, antibodies could block inhibition of adenylyl cyclase by the α_2-adrenergic agonist norepinephrine (Fig. 4). Since we could not identify G_{i1} in platelet membranes, and antibodies EC and QN which bind to G_{i3} and $G_{x(z)}$, respectively, do not block adenylyl cyclase inhibition, we conclude that G_{i2} can couple to the α_2-adrenergic receptor and thereby mediate adenylyl cyclase inhibition (Simonds et al., 1989b). This provides the first evidence in native membranes for interaction between a specific G protein and a receptor-effector.

These results have major implications for our understanding of G protein structure and function, and for the elucidation of the specificity of receptor-effector coupling by G proteins. The ability of carboxy-terminal decapeptide antibodies to block receptor–G protein interaction emphasizes the importance of this domain of the α subunit in receptor interaction. The ability of antibodies to interact with the membrane-bound G protein implies that this region is exposed at the cytoplasmic surface, and is consistent with the ability of pertussis toxin to covalently modify a cysteine within the carboxy-terminal decapeptide in the membrane-bound G protein. Antibody binding to the carboxyterminus, however, does not result in global inhibition of G protein function. Indeed, RM antibody immunoprecipitation of a G protein–effector complex implies that the carboxy-terminal decapeptide of the α subunit is not critically involved in effector interaction.

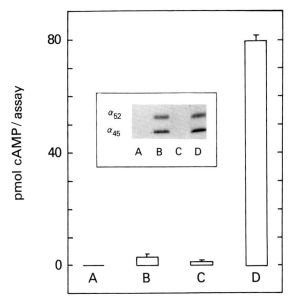

Figure 3. Immunoprecipitation of adenylyl cyclase activity and G_s by RM antibody. Detergent extracts were prepared from bovine brain membranes with (C and D) or without (A and B) preactivation by 100 μM GTP-γ-S for 20 min at 30°C. Extracts (3 mg/ml) were incubated with normal rabbit immunoglobulin (A and C) or RM antibody (B and D) for 4 h at 4°C. Immunoprecipitation was performed with Protein A Staph. Aureus cells (Pansorbin). The immunoprecipitated pellet was washed and resuspended in detergent buffer, and adenylyl cyclase activity was determined. The activity of 20-μl aliquots from each immunoprecipitate is shown. (*Inset*) Parallel immunoblot analysis of α_s content in immunoprecipitates. The positions of the 52- and 45-kD forms of α_s are indicated. (From Simonds et al., 1989c.)

The ability of carboxy-terminal decapeptide antibodies to block receptor–G protein interaction in native membranes should be useful in defining the specificity of these interactions. The studies described above indicate that, at a minimum, one should be able to distinguish coupling to G_{i1} and/or G_{i2} from coupling either to G_{i3} or G_o. Studies are in progress to extend this approach to other tissues and cells with varying proportions of pertussis toxin–sensitive G proteins. Likewise, immunoprecipitation of activated G protein–effector complexes by specific carboxy-terminal peptide antibodies should prove helpful in identifying effectors linked to specific G proteins. Recombinant DNA techniques will clearly provide important information on receptor-effector coupling by G proteins, e.g., α subunits of G_{i1}, G_{i2}, and G_{i3} expressed in *Escherichia coli* all proved capable of stimulating a K$^+$ channel (Yatani et al., 1988), but such approaches can only indicate which G proteins are potentially capable of coupling to individual receptors and effectors. The use of specific antibodies offers the possiblity of identifying the endogenous G protein(s) that actually couple to individual receptors and effectors.

Peptide Antibodies as Probes of G Protein Membrane Attachment

Both G protein α and $\beta\gamma$ are tightly attached to the cytoplasmic surface of the plasma membrane (Gilman, 1987; Spiegel, 1987). Even after activation (and presumed subunit dissociation), neither α nor $\beta\gamma$ subunits are released from the membrane with aqueous buffers lacking detergents (Eide et al., 1987). The molecular basis for the tight association of G protein subunits with the plasma membrane is not apparent from inspection of the primary sequences predicted by cloned cDNAs (Gilman, 1987; Spiegel, 1987; Lochrie and Simon, 1988). None of the subunits show stretches of hydrophobic residues predicted to span the plasma membrane. Hydrodynamic studies of G protein subunits in buffers with and without detergent have led to the suggestion that α subunits are intrinsically hydrophilic, whereas $\beta\gamma$ subunits are hydrophobic and require detergent to prevent aggregation. The failure of α subunits, moreover, to

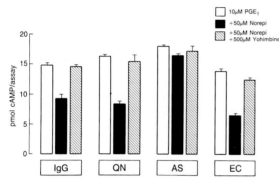

Figure 4. Effect of carboxy-terminal peptide antisera on α_2-adrenergic inhibition of adenylyl cyclase in human platelet membranes. Aliquots of membranes were incubated with control rabbit immunoglobulin (IgG), or affinity-purified antibodies QN, AA, or EC at a final antibody concentration of 50 μg/ml. After 1 h at 4°C, PGE_1-stimulated adenylyl cyclase activity was determined with or without norepinephrine and yohimbine as indicated by the symbols. Values are the mean ± SE of triplicate determinations. (From Simonds et al., 1989b.)

associate with phospholipid vesicles unless these contained $\beta\gamma$ subunits led to the suggestion that the latter serve to anchor α subunits in the plasma membrane (Sternweis, 1986).

Recent studies (Simonds et al., 1989a), involving transient expression of α subunits in COS cells, contradict this idea. Quantitative immunoblotting shows that α_{i3}, for example, can be expressed at levels >10-fold that of β subunits, and yet the expressed α subunits are primarily membrane associated. It is highly likely that at least certain types of α subunit are capable of membrane association independent of $\beta\gamma$ subunits. Earlier studies (Eide et al., 1987) had shown that tryptic cleavage of brain or neutrophil membranes releases α_i and α_0 subunits (minus a 1-2 kD amino-terminal fragment) from the membranes. Thus, α subunits are, in fact, basically cytosolic proteins anchored to the membrane via their amino termini.

Buss et al. (1987) have found that α_i and α_0, but not α_s, are myristylated. Since this cotranslational modification generally involves an amino-terminal glycine residue, it could explain the anchoring role of the amino terminus. To determine the role of myristylation in α subunit membrane association, we expressed α_s and α_{i1} in COS cells, and immunoprecipitated the metabolically labeled products after cell fractionation (Jones et al., 1990). Both α subunits were abundantly expressed and primarily membrane associated, but only α_{i1} was found to incorporate [^{3}H]myristic acid. By site-directed mutagenesis, we altered the second residue of the α_{i1} cDNA from glycine to alanine. The mutant protein failed to incorporate [^{3}H]myristate, and was found primarily in the soluble fraction. Nonetheless, the mutant protein could interact with $\beta\gamma$ subunits, since the latter promoted pertussis toxin–catalyzed ADP-ribosylation of the mutant α subunit. These results indicate that myristylation of α_i, but not α_s, subunits is critical for membrane attachment. The basis for membrane attachement of α_s subunits requires further study.

Both pertussis toxin substrate G protein α subunits, and G protein γ subunits show carboxy-terminal sequence homology with p21 *ras* proteins (Lochrie and Simon, 1988). The latter are known to undergo a complex series of carboxy-terminal posttranslational modifications that are critical for membrane attachment (Schafer et al., 1989). The results with the α_{i1} mutant discussed above make it unlikely that a similar process occurs for α_i subunits, but the possibility that γ subunits undergo *ras*-like carboxy-terminal modifications, and that such modifications are important for membrane attachment of $\beta\gamma$ subunits must be considered.

Acknowledgments

We thank J. Codina, Baylor College of Medicine, for providing G proteins expressed in *Escherichia coli,* and M. Brann, National Institute of Neurological Diseases and Stroke, for collaboration on transient expression of G proteins.

References

Bray, P., A. Carter, C. Simons, V. Guo, C. Puckett, J. Kamholz, A. Spiegel, and M. Nirenberg. 1986. Human cDNA clones for four species of G-α_s signal transduction protein. *Proceedings of the National Academy of Sciences.* 83:8893–8897.

Buss, J. E., S. Mumby, P. Casey, A. Gilman, and B. Sefton. 1987. Myristoylated α subunits of guanine nucleotide regulatory proteins. *Proceedings of the National Academy of Sciences.* 84:7493–7497.

Cerione, R. A., S. Kroll, R. Rajaram, C. Unson, P. Goldsmith, and A. Spiegel. 1988. An antibody directed against the carboxyl-terminal decapeptide of the α subunit of the retinal GTP-binding protein, transducin. *Journal of Biological Chemistry.* 263:9345–9352.

Cerione, R. A., C. Staniszewski, J. Benovic, R. Lefkowitz, M. Caron, P. Gierschik, R. Somers, A. Spiegel, J. Codina, and L. Birnbaumer. 1985. Specificity of the functional interactions of the beta-adrenergic receptor and rhodopsin with guanine nucleotide regulatory proteins reconstituted in phospholipid vesicles. *Journal of Biological Chemistry.* 260:1493–1500.

Eide, B., P. Gierschik, G. Milligan, I. Mullaney, C. Unson, P. Goldsmith, and A. Spiegel. 1987. GTP-binding proteins in brain and neutrophil are tethered to the plasma membrane via their amino termini. *Biochemical and Biophysical Research Communications.* 148:1398–1405.

Fong, H. K. W., K. Yoshimoto, P. Eversole-Cire, and M. Simon. 1988. Identification of a GTP-binding protein alpha subunit that lacks an apparent ADP-ribosylation site for pertussis toxin. *Proceedings of the National Academy of Sciences.* 85:3066–3070.

Gilman, A. 1987. G Proteins: transducers of receptor-generated signals. *Annual Review of Biochemistry.* 56:615–649.

Goldsmith, P., P. Backlund, Jr., K. Rossiter, A. Carter, G. Milligan, C. Unson, and A. Spiegel. 1988a. Purification of heterotrimeric GTP-binding proteins from brain: identification of a novel form of Go. *Biochemistry.* 27:7085–7090.

Goldsmith, P., K. Rossiter, A. Carter, W. Simonds, C. Unson, R. Vinitsky, and A. Spiegel. 1988b. Identification of the GTP-binding protein encoded by Gi3 complementary DNA. *Journal of Biological Chemistry.* 263:6476–6479.

Goldsmith, P., P. Gierschik, G. Milligan, C. Unson, R. Vinitsky, H. Malech, and A. Spiegel. 1987. Antibodies directed against synthetic peptides distinguish between GTP-binding proteins in neutrophil and brain. *Journal of Biological Chemistry.* 262:14683–14688.

Iyengar, R., and L. Birnbaumer. 1987. Signal transduction by G-proteins. *ISI Atlas of Science: Pharmacology.* 1:213–221.

Jones, D. T., and R. Reed. 1989. G_{olf}: an olfactory neuron-specific G protein involved in odorant signal transduction. *Science.* 244:790–795.

Jones, T. L. Z., W. F. Simonds, J. J. Merendino, Jr., M. R. Brann, and A. M. Spiegel. 1990. Myristylation of an inhibitory GTP-binding protein alpha subunit is essential for its membrane attachment. *Proceedings of the National Academy of Sciences.* 87:568–572.

Lochrie, M.A., and M. Simon. 1988. G protein multiplicity in eukaryotic signal transduction systems. *Biochemistry.* 27:4957–4965.

Matsuoka, M., H Itoh, K. Tohru, and Y. Kaziro. 1988. Sequence analysis of cDNA and genomic DNA for a putative pertussis toxin-insensitive guanine nucleotide-binding regulatory protein α subunit. *Proceedings of the National Academy of Sciences.* 85:5384–5388.

Roof, D. J., M. Applebury, and P. Sternweis. 1985. Relationships within the family of GTP-binding proteins isolated from bovine central nervous system. *Journal of Biological Chemistry.* 260:16242–16249.

Schafer, W. R., R. Kim, R. Sterne, J. Thorner, S.-H. Kim, and J. Rine. 1989. Genetic and pharmacological suppression of oncogenic mutations in *ras* genes of yeast and humans. *Science.* 245:379–385.

Senogles, S. E., J. Benovic, N. Amlaiky, C. Unson, G. Milligan, R. Vinitsky, A. Spiegel, and M. Caron. 1987. The D2-dopamine receptor of anterior pituitary is functionally associated with a pertussis toxin–sensitive guanine nucleotide binding protein. *Journal of Biological Chemistry.* 262:4860–4867.

Simonds, W. F., R. M. Collins, A. M. Spiegel, M. R. Brann. 1989a. Membrane attachment of recombinant G-protein alpha subunits in excess of beta/gamma subunits in a eukaryotic expression system. *Biochemical and Biophysical Research Communications.* 164:46–53.

Simonds, W. F., P. Goldsmith, J Codina, C. Unson, and A. Spiegel. 1989b. G_{i2} mediates α_2-adrenergic inhibition of adenylyl cyclase in platelet members: in situ identification with $G\alpha$ C-terminal antibodies. *Proceedings of the National Academy of Sciences.* 86:7809–7813.

Simonds, W. F., P. Goldsmith, C. Woodard, C. Unson, and A. Spiegel. 1989c. Receptor and effector interactions of G_s. *Federation of European Biochemical Societies Letters.* 249:189–194.

Spiegel, A. 1987. Signal transduction by guanine nucleotide binding proteins. *Molecular and Cellular Endocrinology.* 49:1–16.

Sternweis, P.C. 1986. The purified α subunits of G_o and G_i from bovine brain require $\beta\gamma$ for association with phsopholipid vesicles. *Journal of Biological Chemistry.* 261:631–637.

Sullivan, K. A., R. Miller, S. Masters, B. Beiderman, W. Heideman, and H. Bourne. 1987. Identification of receptor contract site involved in receptor-G protein coupling. *Nature*. 330:758–760

West, R. E., Jr., J. Moss, M. Vaughan, T. Liu, and T.-Y. Liu. 1985. Pertussis toxin-catalyzed ADP-ribosylation of transducin: cysteine 347 is the ADP-ribose acceptor site. *Journal of Biological Chemistry*. 260:14428–14430.

Yatani, A., R. Mattera, J. Codina, R. Graf, K. Okabe, E. Padrell, R. Iyengar, A. Brown, and L. Birnbaumer. 1988. The G protein-gated atrial K^+ channel is stimulated by three distinct G_i α-subunits. *Nature*. 336:680–682.

Chapter 17

Mechanisms of G Protein Action: Insight from Reconstitution

Paul C. Sternweis and Iok-Hou Pang

Department of Pharmacology, University of Texas, Southwestern Medical Center, Dallas, Texas 75235

Introduction

The GTP-dependent regulatory proteins (G proteins) provide signaling pathways for many extracellular stimuli. These proteins were first identified as mediators of hormonal regulation of adenylyl cyclase (G_s and G_i proteins) and of light-dependent stimulation of cGMP-specific phosphodiesterase in mammalian retina (G_t proteins). More recently, the G proteins have also been implicated in the regulation of various phospholipases and ion channels. The structures and functions of this family of proteins have been discussed in several reviews (Gilman, 1987; Casey et al., 1988; Lochrie and Simon, 1988; Neer and Clapham, 1988; Ross, 1989).

The term, G protein, has been used to describe any protein that binds guanine nucleotides with high specificity. In this discussion, G protein will refer only to those GTP-binding proteins that are composed of three subunits (α, β, and γ) and are known to interact with membrane-associated receptors. In the two well-defined systems, G_s and G_t proteins have been shown to directly activate enzymes (adenylyl cyclase and cGMP-phosphodiesterase) that regulate concentrations of second messenger molecules (cAMP and cGMP). This has given rise to the simple generalization that these pathways consist of three components: receptors, G proteins, and effector molecules.

$$\text{Recepter} \rightarrow \text{G protein} \rightarrow \text{Effector}$$

Clearly, the effector can give rise to extended pathways of regulation. Furthermore, some activities currently identified as effectors (such as the phospholipases and ion channels) may be several steps removed from the G proteins.

Individual G proteins are identified by their α subunits. These subunits appear to be most diverse, contain the binding site for guanine nucleotides, and can be ADP-ribosylated by either cholera toxin (G_s and G_t proteins) or pertussis toxin (G_i, G_o, and G_t proteins). An exception to the latter property is a putative α subunit identified by cloning that is highly homologous to the α_i proteins but lacks the site for modification by pertussis toxin (Fong et al., 1988; Matsuoka et al., 1988). The β and γ subunits appear to be less numerous. They are always found as a tightly associated complex and can associate interchangeably with various α subunits.

Activation of G proteins is achieved by exchanging bound GDP for GTP. This is accompanied by dissociation of the α(GTP) from $\beta\gamma$ subunits and thus the production of two potential regulatory species. The α subunits contain an endogenous GTPase activity. The relatively slow hydrolysis of bound GTP can return the α subunit to its basal state. This facilitates reassociation with the $\beta\gamma$ subunits. Fig. 1 is a schematic for the potential steps involved in this cycle.

The following discussion will examine results that contribute to three areas in G protein regulation: first, the interaction of G proteins with receptors (Florio and Sternweis, 1985, 1989); second, isolation of α subunits by affinity with immobilized $\beta\gamma$ (Pang and Sternweis, 1989); third, identification of G protein pathways used in the regulation of Ca^{++} currents by neuropeptides (Ewald et al., 1988, 1989).

Receptor Interaction with G Proteins

Reconstitution paradigms with purified proteins in several laboratories indicated that only the receptor, a single polypeptide, and the G protein were sufficient to observe agonist stimulation of guanine nucleotide exchange. Are all of the subunits of the G protein required? The necessity of the α subunit was clear as it contains the binding site for guanine nucleotides. A role for the $\beta\gamma$ subunits was less obvious. Their requirement for efficient stimulation of guanine nucleotide exchange has been demonstrated in two experimental systems (Fung, 1983; Phillips and Cerione, 1988; Florio and Sternweis, 1989). Furthermore, the $\beta\gamma$ subunits were also required for the expression of guanine nucleotide–sensitive high-affinity binding of agonists to muscarinic receptors (Florio and Sternweis, 1989). It thus appears that the heterotrimeric form of the G protein interacts best with the receptor. It is possible that $\beta\gamma$ may only facilitate interaction of the α subunit with the receptor; for example, it may help localize α subunits to the vicinity of the receptor (Sternweis, 1986). Alternately, the role of $\beta\gamma$ may be much greater and a diversity in $\beta\gamma$ species may provide a mechanism for events such as differential recognition by receptors.

Activation of G proteins is effected by exchange of guanine nucleotides (GTP for GDP); this exchange (and thus activation) can happen spontaneously. The rate of exchange will depend on the rate of dissociation of GDP and the availability of GTP. Inactivation of the G protein depends on the rates of GTP hydrolysis and dissociation. It is likely that the rate of dissociation of GTP in the presence of Mg^{++} is insignificant relative to that of hydrolysis. Therefore, in the presence of saturating GTP, the ratio of

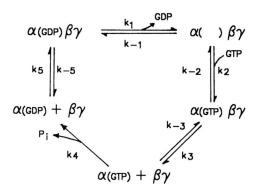

Figure 1. Abbreviated scheme for guanine nucleotide exchange and activation of G proteins. Additional pathways clearly exist (such as dissociation of subunits before or during exchange) but are not included for the purpose of this discussion.

activated to nonactivated G protein will be the ratio of the rates of GDP dissociation and GTP hydrolysis (k_1/k_4 in Fig. 1). Without stimulation, this ratio is low.

The current widely accepted model for activation by receptors focuses on the dissociation of GDP. This is assumed to be the rate-limiting step for activation; receptors stimulate this rate. Early experiments demonstrated that agonists could accelerate the rate of release of both GDP and the GTP analogue, Gpp(NH)p (Cassel and Selinger, 1977a, b). Thus, the receptor is proposed to "open up" the guanine nucleotide-binding site on the G proteins. After release of GDP, the unoccupied G protein rapidly binds GTP, which is assumed to be the most abundant form of the nucleotide.

But, what if GDP is more abundant? In this case, association of GDP (k_{-1}, Fig. 1) would be significant and activation by the receptor, as well as basal activation would be dampened. This could be avoided if the receptor also had an effect on the relative association rates for GTP and GDP. Evidence for such an effect was observed during reconstitution of purified muscarinic receptors with G_o proteins in phospholipid vesicles (Florio and Sternweis, 1989). Initial experiments that attempted to demonstrate muscarinic stimulation of guanine nucleotide exchange were compromised by the rapid basal exchange rates of the G_o protein and relatively poor coupling. Thus, stimulation of GTP γS binding (an essentially irreversible reaction under the conditions used) by the agonist oxotremorine was very small. When GDP was added to the reaction at 50-fold higher concentrations than GTPγS, the rate of GTPγS binding in the absence of agonist was markedly reduced while the stimulation of binding by agonist was retained and became proportionally larger. If GTP was used instead of GDP, both the basal- and agonist-stimulated binding of GTPγS were reduced. Finally, the action of the receptor was shown to alter the ratio of GTP:GDP that could be bound to G_o. When equal concentrations of the nucleotides were mixed with the vesicles, binding of GDP was favored in the absence of agonist, while GTP was favored in its presence. The latter experiment actually examined initial binding to nucleotide free G protein. These data therefore suggest that the receptor can alter the relative association rates of the G protein for GDP and GTP. Thus according to Fig. 1, receptors either reduce k_{-1} with respect to k_2 or stimulate k_2 with respect to k_{-1}. The eventual delineation of the specific mechanism awaits a well-coupled system better suited to the kinetic measurements.

What is the importance of such an effect? The key action of the receptor is still the acceleration of dissociation of GDP. The relative impact of this stimulation will depend on the efficiency of exchange. This efficiency would be guaranteed to be high if receptors altered the rates of association such that GTP was highly favored over GDP. When GTP is the only available replacement for binding, there is a small effect. However, the concentration of GDP in the vicinity of the G proteins could be quite high, especially during agonist-induced rapid turnover of G proteins. In this case, the effect would be large and ensure the binding of GTP during stimulation by receptors.

Isolation and Identification of G Protein α Subunits

Recent experimentation in this laboratory has focused on the preparation of specific α subunits for use in the exploration of their functions. To enhance these efforts, we have developed an affinity matrix consisting of immobilized βγ subunits (Pang and Sternweis, 1989). This method takes advantage of the apparent relative lack of specificity of the βγ subunits for various α subunits and the capability of controlling α and βγ subunit interaction in solution. So far, we have found this matrix useful for the final isolation of α subunits from each other and the isolation of α subunits from crude preparations in one step.

Synthesis of the affinity matrix took advantage of the relative insensitivity of βγ to sulfhydryl reagents. The cross-linking reagent, m-maleimidobenzoylsulfosuccinimide ester (Pierce Chemical Co., Rockford, IL) was used to couple the amino group of ω-aminobutyl agarose (Sigma Chemical Co., St. Louis, MO) to sulfhydryl groups on the βγ (Pang and Sternweis, 1989). In this fashion, matrices were synthesized that contain 3–7 nmol of binding sites for α subunits. This represents ~25% of the potential binding sites predicted from the quantity of immobilized βγ subunits. The immobilized βγ has an affinity for α subunits similar to that of free βγ. Thus, the EC_{50} for inhibiting binding of $α_o$ to a matrix that contained 3 μM binding sites was 2 μM free βγ.

One of the problems in purifying α subunits from brain is the huge abundance of the $α_o$ protein. Thus the two $α_i$ proteins that could be isolated from brain were always "contaminated" with $α_o$. Final isolation of the $α_{i1}$ and $α_{i2}$ proteins could be achieved by taking advantage of the apparent higher affinity of βγ for $α_i$ proteins (Sternweis,

1986). When a mixture of α_i and α_o were associated with the matrix, the bulk of the α_o could be eluted with milder activating conditions (GTP and NaCl). Addition of Mg^{++} and the activating ligand AlF_4^- (Sternweis and Gilman, 1982) completed the elution of α_o and started elution of the α_i. Subsequent increases in Mg^{++} concentration then yielded the remainder of α_i in a pure form. In addition to the separation of different α subunits, the procedure also removes other contaminants in the preparations. This is accomplished in two ways. First, the $\beta\gamma$ selects for binding α subunits. Second, the elution is specific in that conditions are used which activate the G proteins and promote dissociation of the subunits; nonspecifically associated proteins are left bound to the matrix.

The selectivity of the matrix made it possible to isolate α subunits directly from extracts of membrane preparations. When bovine brain membranes were examined, three polypeptides were specifically isolated. They comigrated with the α_{i1} (41 kD), α_{i2} (40 kD), and α_o (39 kD) proteins that were previously purified from this source (Sternweis and Robishaw, 1984; Mumby et al., 1988). Subsequent analysis with subunit-specific antibodies (Pang and Sternweis, 1989) confirmed the presence of these three α subunits and further demonstrated that the matrix was capable of isolating at least 50% of the α_o (weakest association with $\beta\gamma$) and >80% of the α_i subunits from these extracts. The isolation of α_s could also be easily examined by its ability to stimulate adenylyl cyclase; 80% of this activity was isolated by the procedure. Thus, it appears that isolation of α subunits from crude extracts can produce a clear picture of the content of known α subunits in thes preparations.

An example of such an analysis is shown in Fig. 2. Here, solubilized membranes from rat liver were examined. Three polypeptides of 40–42 kD were eluted specifically by AlF_4^-. The polypeptide that migrates with a molecular weight of 40,000 probably represents α_{i2} (previously identified with antibodies). The nature of the other two proteins is less clear. However, they are apparently different from α_{i1}, since they do not comigrate with the purified α subunit. Furthermore, α_{i1} was not readily detectable in liver with a specific antiserum (Mumby et al., 1988). One of the larger polypeptides could be α_{i3} which migrates with a mobility similar to α_{i1} in this system. An intriguing and likely possibility is that the other, or perhaps both, larger proteins may represent new species of α subunits of G proteins. If so, one of the goals of this matrix, its use to identify new α subunits, will be fulfilled.

G Protein Pathways in the Regulation of Calcium Currents

The purified α subunits can be used to help identify the G protein pathways used for regulation of various events. Most cells contain several G proteins and these proteins regulate many pathways. A current problem facing the field is one of circuitry. Which receptors interact with which G proteins and which G proteins regulate which effector pathways? The only two clearly defined pathways are the G_s regulation of adenylyl cyclase and the G_t regulation of cGMP phosphodiesterase. No other G protein has been shown to stimulate these enzymes. Recent experiments that examine the regulation of ion channels give quite a different picture. At least four G proteins may be capable of regulating K^+ channels in atrial cells (Logothetis et al., 1988; VanDongen et al., 1988; Yatani et al., 1988). Much of these data, experiments examining regulation of K^+ currents in other cells, and the question of which subunits support this regulation are presented in other papers contained in this volume.

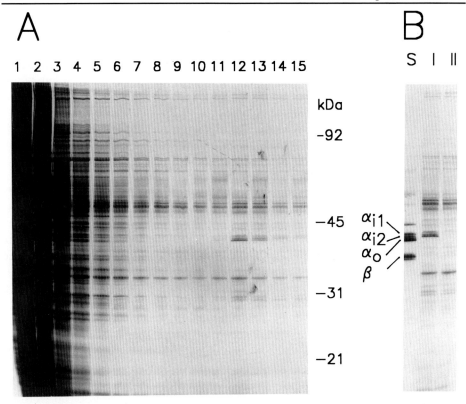

Figure 2. Isolation of α subunits from the membrane of rat liver by $\beta\gamma$ affinity chromatography. (*A*) A cholate extract of liver membranes (0.5 ml containing 2 mg protein and 0.4 nmol GTPγS binding site) was incubated with 0.5 ml of $\beta\gamma$-agarose (containing 3 nmol of α subunit binding sites) as described (Pang and Sternweis, 1989). The matrix was washed (fractions *1–10*, 0.5 ml each fraction) and then eluted with AlF$_4^-$ (fractions *11–15*). In this example, the eluate contained 11% of the GTPγS binding sites and 66% of the total G$_s$ activity loaded onto the $\beta\gamma$-agarose. Proteins from each fraction were separated by SDS-PAGE and visualized with silver staining. Locations of molecular weight standards are shown on the right. (*B*) The membrane extracts were treated with either an active or a heat-inactivated $\beta\gamma$-agarose column. The protein contents in the second fraction of the AlF$_4^-$ eluate (fraction *12*) from either column were analyzed by SDS-PAGE and stained with silver (*S*, a standard preparation containing α_{i1}, α_{i2}, α_o, and β from bovine brain; *I*, eluate from active $\beta\gamma$-agarose; *II*, eluate from heated $\beta\gamma$-agarose).

This discussion will confine itself to experiments performed by Ewald and colleagues (Ewald et al., 1988, 1989) on the inhibitory regulation of voltage-activated Ca^{++} currents by neuropeptide Y (NPY) and bradykinin (BK) in dorsal root ganglion (DRG) cells of the rat. In the preparations of cells utilized, this inhibition of both the transient and sustained currents by the neuropeptides is attenuated by pertussis toxin. This allowed the experimental paradigm in which specific purified α subunits could be introduced into treated cells in attempts to replace modified and inactive α subunits. In the first experiments (Ewald et al., 1988), the purified α_o was shown to restore inhibition of the Ca^{++} currents by NPY. The GDP form of the α subunit was introduced into the cell by way of the patch pipette used to voltage clamp the cell.

Restoration was time dependent and did not occur if the subunit was heat inactivated prior to use. When the purified α_{i1} and α_{i2} subunits were used, the restoration of NPY regulation was less or did not occur, respectively. This suggested that the main pathway for inhibiting these channels was via the G_o protein. This would be similar to the results reported for NG108 cells in which G_o was found to be better than G_i for restoring inhibition of calcium currents by opioids (Hescheler et al., 1987; Rosenthal et al., 1988).

The correct interpretation for this specificity followed from subsequent experiments. In contrast to NPY, restoration of inhibition by BK could be effected by all three α subunits (Ewald et al., 1989). In the same cell reconstituted with α_{i2}, regulation by BK, but not NPY, was restored. Thus multiple G protein pathways are capable of modulating the calcium currents. The apparent specificity of NPY for G_o presumably reflects the selectivity of the NPY receptor for this G protein rather than a specific regulation of the ion channels.

A second interesting observation from these experiments was the lack of complete restoration of BK regulation by any single α subunit. Full restoration was achieved with a combination of the α_o and α_{i2} subunits. This suggests that multiple pathways may be used to achieve the full effects of a single stimulus. At this point, it is not known whether this regulation by BK is derived from a single receptor subtype lacking

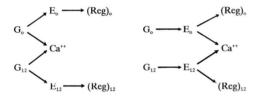

Figure 3. Potential pathways for the observed regulation of Ca^{++} currents and other events in DRG cells by G proteins. Whether the regulation of Ca^{++} channels by G proteins is more direct (*left*) or through second messenger pathways (*right*) is unknown.

specificity for G proteins or through several receptor subtypes acting on specific G proteins.

What might all this mean? Why should receptors use different G proteins to achieve the same end result? It is possible that the inhibition of Ca^{++} currents in these cells is a common point of regulation by many neuromodulators but that other regulated events are more specific (Fig. 3). Thus, NPY regulates Ca^{++} and other events that are regulated specifically by G_o, $(Eff)_o$. BK may then regulate a broader spectrum of events, such that it would inhibit Ca^{++} but also effect modulation of activities regulated by several G proteins, $(Eff)_o$, and $(Eff)_i$. Whether all of this regulation occurs throughout the whole cell or is more localized will be a question of future study.

Conclusion

The G protein story is just beginning. We have just examined three small areas. These experiments have provided some insight into the mechanism of G protein pathways. Yet, they are equally notable for demonstraing some of the questions that remain. Thus, it is clear that many of the details of the mechanism of receptor activation of G proteins remain to be determined. While certain key events that describe the general picture have been shown, the kinetic details may provide further insight into much of the action that may be operative under more physiological conditions.

The proliferation of known G proteins has escalated the question of their specificity of interaction with various receptors and effector pathways. This proliferation will probably continue. We have described one technique ($\beta\gamma$ affinity chromatography) that should aid in the search to define the players. This technique may also help provide clues to specificity by determining which G proteins are present in a given system.

A second approach to specificity is to intervene in a system with purified proteins to find out which can restore, amplify, or inhibit a designated signal. Experiments using reconstitution with purified proteins to study regulation of ion channels have provided some answers, but the answers raise new complications. Thus, these results have suggested the possibility that an "effector" can be regulated by several G proteins, rather than a singular pathway.

It is probable that the regulation of many effector pathways by G proteins will not be as simple as previously thought. The network may be complex and interactive. While more players (receptors, G proteins, second messengers) remain to be found, the current field alone suggests that our complete understanding of this regulation will remain elusive for some time.

Acknowledgments

This work was supported by National Institutes of Health Grant GM-31954 and grants from The Searle Scholars Program and the March of Dimes. Paul C. Sternweis is an Established Investigator of the American Heart Association.

References

Casey, P. J., M. P. Graziano, M. Freissmuth, and A. G. Gilman. 1988. Role of G Proteins in Transmembrane Signaling. *In* Cold Spring Harbor Symposia on Quantitative Biology. LIII:203–208.

Cassel, D., and Z. Selinger. 1977a. Catecholamine-induced release of [^{3}H]-G_{pp}(NH)p from turkey erythrocyte adenylate cyclase. *Journal of Cyclic Nucleotide Research.* 3:11–22.

Cassel, D., and Z. Selinger. 1977b. Mechanism of adenylate cyclase activation by cholera toxin: an inhibition of GTP hydrolysis at the regulatory site. *Proceedings of the National Academy of Sciences.* 74:3307–3311.

Ewald, D. A., I.-H. Pang, P. C. Sternweis, and R. J. Miller. 1989. Differential G protein-mediated coupling of neurotransmitter receptors to calcium channels in rat dorsal root ganglion neurons in vitro. *Neuron.* 2:1185–1193.

Ewald, D. A., P. C. Sternweis, and R. J. Miller. 1988. Guanine nucleotide-binding protein Go-induced coupling of neuropeptide Y receptors to Ca^{++} channels in sensory neurons. *Proceedings of the National Academy of Sciences.* 85:3633–3637.

Florio, V. A., and P. C. Sternweis. 1985. Reconstitution of resolved muscarinic cholinergic receptors with purified GTP-binding proteins. *Journal of Biological Chemistry.* 260:3477–3483.

Florio, V. A., and P. C. Sternweis. 1989. Mechanisms of muscarinic receptor action on G_o in reconstituted phospholipid vesicles. *Journal of Biological Chemistry.* 264:3909–3915.

Fong, H. K. W., K. K. Yoshimoto, P. Eversole-Cire, and M. I. Simon. 1988. Identification of a GTP-binding protein alpha subunit that lacks an apparent ADP-ribosylation site for pertussis toxin. *Proceedings of the National Academy of Sciences.* 85:3066–3070.

Fung, B. K.-K. 1983. Characterization of transducin from bovine retinal rod outer segments. I. Separation and reconstitution of subunits. *Journal of Biological Chemistry*. 258:10495–10502.

Gilman, A. G. 1987. G proteins: transducers of receptor-generated signals. *Annual Review of Biochemistry*. 56:615–649.

Hescheler, J., W. Rosenthal, W. Trautwein, and G. Schultz. 1987. The GTP-binding protein, G_o, regulates neuronal calcium channels. *Nature*. 325:445–447.

Lochrie, M. A., and M. I. Simon. 1988. G protein multiplicity in eukaryotic signal transduction systems. *Biochemistry*. 27:4957–4965.

Logothetis, D. E., D. Kim, J. K. Northup, E. J. Neer, and D. E. Clapham. 1988. Specificity of action of guanine nucleotide-binding regulatory protein subunits on the cardiac muscarinic K^+ channel. *Proceedings of the National Academy of Sciences*. 85:5814–5818.

Matsuoka, M., H. Itoh, T. Kozasa, and Y. Kaziro. 1988. Sequence analysis of cDNA and genomic DNA for a putative pertussis toxin-insensitive guanine nucleotide-binding regulatory protein alpha subunit. *Proceedings of the National Academy of Sciences*. 85:5384–5388.

Mumby, S., I.-H. Pang, A. G. Gilman, and P. C. Sternweis. 1988. Chromatographic resolution and immunologic identification of the alpha 40 and alpha 41 subunits of guanine nucleotide-binding regulatory proteins from bovine brain. *Journal of Biological Chemistry*. 263:2020–2026.

Neer, E. J., and D. E. Clapham. 1988. Roles of G protein subunits in transmembrane signalling. *Nature*. 333:129–134.

Pang, I.-H., and P. C. Sternweis. 1989. Isolation of the alpha subunits of GTP-binding regulatory proteins by affinity chromatography with immobilized beta-gamma subunits. *Proceedings of the National Academy of Sciences*. 86:7814–7818.

Phillips, W. J., and R. A. Cerione. 1988. The intrinsic fluorescence of the α subunit of transducin. *Journal of Biological Chemistry*. 263:15498–15505.

Rosenthal, W., J. Hescheler, W. Trautwein, and G. Schultz. 1988. Control of voltage-dependent calcium channels by G protein-coupled receptors. *FASEB Journal*. 2:2784–2790.

Ross, E. M. 1989. Signal sorting and amplification through G protein-coupled receptors. *Neuron*. 3:141–152.

Sternweis, P. C. 1986. The purified α subunits of G_o and G_i from bovine brain require $\beta\gamma$ for association with phospholipid vesicles. *Journal of Biological Chemistry*. 261:631–637.

Sternweis, P. C., and A. G. Gilman. 1982. Aluminum: a requirement for activation of the regulatory component of adenylate cyclase by fluoride. *Proceedings of the National Academy of Sciences*. 79:4888–4891.

Sternweis, P. C., and J. D. Robishaw. 1984. Isolation of two proteins with high affinity for guanine nucleotides from membranes of bovine brain. *Journal of Biological Chemistry*. 259:13806–13813.

VanDongen, A. M. J., J. Codina, J. Olate, R. Mattera, R. Joho, L. Birnbaumer, and A. M. Brown. 1988. Newly identified brain potassium channels gated by the guanine nucleotide binding protein G_o. *Science*. 242:1433–1437.

Yatani, A., R. Mattera, J. Codina, R. Graf, K. Okabe, E. Padrell, R. Iyengar, A. M. Brown, and L. Birnbaumer. 1988. The G protein-gated atrial K^+ channel is stimulated by three distinct G_i-alpha subunits. *Nature*. 336:680–682.

List of Contributors

Jeffrey L. Benovic, Fels Institute for Cancer Research and Molecular Biology, Temple University School of Medicine, Philadelphia, Pennsylvania

Laurent Bernheim, Department of Physiology and Biophysics, University of Washington School of Medicine, Seattle, Washington

Lutz Birnbaumer, Department of Molecular Physiology and Biophysics, and the Department of Cell Biology, Baylor College of Medicine, Houston, Texas

Martha M. Bosma, Department of Physiology and Biophysics, University of Washington School of Medicine, Seattle, Washington

Jose L. Boyer, Department of Pharmacology, University of North Carolina School of Medicine, Chapel Hill, North Carolina

Mark R. Brann, Laboratory of Molecular Biology, National Institute of Neurological Disorders and Stroke, Bethesda, Maryland

Arthur M. Brown, Department of Molecular Physiology and Biophysics, Baylor College of Medicine, Houston, Texas

Marc G. Caron, Howard Hughes Medical Institute, Duke University Medical Center, Durham, North Carolina

David E. Clapham, Departments of Pharmacology and of Physiology and Biophysics, Mayo Foundation, Rochester, Minnesota

Juan Codina, Department of Cell Biology, Baylor College of Medicine, Houston, Texas

Sunita de Sousa, Howard Hughes Medical Institute and the Department of Biochemistry, University of Washington, Seattle, Washington

Peter N. Devreotes, Department of Biological Chemistry, The Johns Hopkins University School of Medicine, Baltimore, Maryland

N. Dhanasekaran, Division of Basic Sciences, National Jewish Center for Immunology and Respiratory Medicine, Denver, Colorado

Richard A. F. Dixon, Department of Molecular Biology, Merck, Sharp and Dohme Research Laboratories, West Point, Pennsylvania

A. C. Dolphin, Department of Pharmacology, St. George's Hospital Medical School, London SW17 ORE, United Kingdom

C. Peter Downes, Department of Pharmacology, University of North Carolina School of Medicine, Chapel Hill, North Carolina

Victor M. Garsky, Department of Molecular Biology, Merck, Sharp and Dohme Research Laboratories, West Point, Pennsylvania

Jackson B. Gibbs, Department of Molecular Biology, Merck, Sharp and Dohme Research Laboratories, West Point, Pennsylvania

Paul K. Goldsmith, Molecular Pathophysiology Branch, National Institute of Diabetes, Digestive, and Kidney Diseases, National Institutes of Health, Bethesda, Maryland

Rolf Graf, Department of Cell Biology and Molecular Physiology and Biophysics, Baylor College of Medicine, Houston, Texas

Robert E. Gunderson, Department of Biological Chemistry, The Johns Hopkins University School of Medicine, Baltimore, Maryland

Sunil K. Gupta, Division of Basic Sciences, National Jewish Center for Immunology and Respiratory Medicine, Denver, Colorado

T. Kendall Harden, Department of Pharmacology, University of North Carolina School of Medicine, Chapel Hill, North Carolina

Lynn E. Heasley, Division of Basic Sciences, National Jewish Center for Immunology and Respiratory Medicine, Denver, Colorado

John R. Hepler, Department of Pharmacology, University of North Carolina School of Medicine, Chapel Hill, North Carolina

Bertil Hille, Deparment of Physiology and Biophysics, University of Washington School of Medicine, Seattle, Washington

James B. Hurley, Howard Hughes Medical Institute and the Department of Biochemistry, University of Washington, Seattle, Washington

Hiroyuki Ito, The Second Department of Internal Medicine, Faculty of Medicine, University of Tokyo, Hongo, Bunkyo-ku, Tokyo 113, Japan

Gary L. Johnson, Division of Basic Sciences, National Jewish Center for Immunology and Respiratory Medicine, Denver, Colorado

S. V. Penelope Jones, Laboratory of Molecular Biology, National Institute of Neurological Disorders and Stroke, Receptor Genetics, Inc., Bethesda, Maryland

Teresa L. Z. Jones, Molecular Pathophysiology Branch, National Institute of Diabetes, Digestive, and Kidney Diseases, National Institutes of Health, Bethesda, Maryland

Glenn E. Kirsch, Department of Molecular Physiology and Biophysics, Baylor College of Medicine, Houston, Texas

Yoshihisa Kurachi, The Second Department of Internal Medicine, Faculty of Medicine, University of Tokyo, Hongo, Bunkyo-ku, Tokyo 113, Japan

Robert J. Lefkowitz, Howard Hughes Medical Institute, Duke University Medical Center, Durham, North Carolina

Mark D. Leibowitz, Merck, Sharp and Dohme Research Laboratories, Rahway, New Jersey

Pamela Lilly, Department of Biological Chemistry, The Johns Hopkins University School of Medicine, Baltimore, Maryland

Mark S. Marshall, Department of Molecular Biology, Merck, Sharp and Dohme Research Laboratories, West Point, Pennsylvania

Rafael Mattera, Departments of Cell Biology and Molecular Physiology and Biophysics, Baylor College of Medicine, Houston, Texas

Andrew J. Morris, Department of Pharmacology, University of North Carolina School of Medicine, Chapel Hill, North Carolina

Chaya Nanavati, Departments of Pharmacology and of Physiology and Biophysics, Mayo Foundation, Rochester, Minnesota

Neil M. Nathanson, Department of Pharmacology, University of Washington, Seattle, Washington

List of Contributors

Eva J. Neer, Department of Medicine, Brigham and Women's Hospital and Harvard Medical School, Boston, Massachusetts

Koji Okabe, Departments of Cell Biology and Molecular Physiology and Biophysics, Baylor College of Medicine, Houston, Texas

James J. Onorato, Howard Hughes Medical Institute, Duke University Medical Center, Durham, North Carolina

Shoji Osawa, Division of Basic Sciences, National Jewish Center for Immunology and Respiratory Medicine, Denver, Colorado

Paul J. Pfaffinger, Center for Neurobiology and Behavior, Columbia University CPS, Research Annex, New York, New York

Iok-Hou Pang, Department of Pharmacology, University of Texas, Southwestern Medical Center, Dallas, Texas

Geoffrey S. Pitt, Department of Biological Chemistry, The Johns Hopkins University School of Medicine, Baltimore, Maryland

Nicole Provost, Howard Hughes Medical Institute and Department of Biochemistry, University of Washington, Seattle, Washington

Maureen B. Pupillo, Department of Biological Chemistry, The Johns Hopkins University School of Medicine, Baltimore, Maryland

Randall R. Reed, Howard Hughes Medical Institute, Department of Molecular Biology and Genetics, The Johns Hopkins University School of Medicine, Baltimore, Maryland

Michael D. Schaber, Department of Molecular Biology, Merck, Sharp, and Dohme Research Laboratories, West Point, Pennsylvania

Edward M. Scolnick, Department of Molecular Biology, Merck, Sharp and Dohme Research Laboratories, West Point, Pennsylvania

R. H. Scott, Department of Pharmacology, St. George's Hospital Medical School, London SW17 0RE, United Kingdom

William F. Simonds, Molecular Pathophysiology Branch, National Institute of Diabetes, Digestive, and Kidney Diseases, National Institute of Health, Bethesda, Maryland

Allen M. Spiegel, Molecular Pathophysiology Branch, National Institute of Diabetes, Digestive, and Kidney Diseases, National Institutes of Health, Bethesda, Maryland

Paul C. Sternweis, Department of Pharmacology, University of Texas, Southwestern Medical Center, Dallas, Texas

Cecilia G. Unson, Department of Biochemistry, Rockefeller University, New York, New York

Antonius M. J. VanDongen, Department of Molecular Physiology and Biophysics, Baylor College of Medicine, Houston, Texas

Roxanne A. Vaughan, Department of Biological Chemistry, The Johns Hopkins University School of Medicine, Baltimore, Maryland

Ursula S. Vogel, Department of Molecular Biology, Merck, Sharp and Dohme Research Laboratories, West Point, Pennsylvania

Gary L. Waldo, Department of Pharmacology, University of North Carolina School of Medicine, Chapel Hill, North Carolina

Jürgen Wess, Laboratory of Molecular Biology, National Institute of Neurological Disorders and Stroke, Bethesda, Maryland

Stuart Yarfitz, Howard Hughes Medical Institute and Department of Biochemistry, University of Washington, Seattle, Washington

Atsuko Yatani, Department of Molecular Physiology and Biophysics, Baylor College of Medicine, Houston, Texas

Subject Index

α subunits, 117, 143
Adenylyl cyclase, 77, 133, 153, 185
Affinity chromatography, 197

$\beta\gamma$ dimers, 169
Bradykinin, 197
Brain G protein, 143

CHO cells, 133
Calcium,
　channels, 11
　currents, 11, 197
　signaling, 43
cAMP levels, 105
Cancer, 77
Cardiac electrophysiology, 29
Cardiac muscarinic K channel, 29
Chemotaxis, 125
Chick, 133
Cloning, 105
Constitutive active, 117
Coupling to G proteins, 105
Cytosol-localized Ras mutant, 77

Desensitization, 87
　of receptors, 43
Dictyostelium, 125
Direct gating, 1
Dorsal root ganglion neurons, 11
Drosophila, 133, 143, 157

Effectors, 185
Embryonic development, 157

G proteins, 1, 11, 29, 61, 117, 125, 143, 153, 157, 169, 185, 197
G_{olf}, 153
GTPase, 169
GTPase-activating protein, 77
GTP-binding proteins, 29
GTP-dependent regulatory proteins, 197

Heart, 133,
　G protein, 143

In situ hybridization, 157
Indirect gating, 1
Ion channels, 1, 29, 105, 169, 197

M current, 43
mRNA distribution, 105
Muscarinic acetylcholine receptor, 133
Muscarinic receptor, 197
Mutation, 117

N channels, 11
Networks, 1
Neuropeptide Y, 197

Olfaction, 153
Olfactory receptor, 153
Oncogene, 77
Ovarian development, 157

Patch clamp, 29
Peptide,
　antibodies, 185
　neurotransmitters, 43
Pertussis toxin, 185
Phosphoinositide,
　metabolism, 105
　signaling, 61
Phospholipase C, 61, 133
Phosphorylation, 87
Protein kinase C, 43
Protein kinases, 87
Protein turnover, 143

Ras antagonists, peptides, 77
Receptor coupling, 29
Receptors, 87, 185
Reconstitution, 197

Saccharomyces cerevisiae, 77
Signal transduction, 29, 125, 169, 185
Structure/function, 77
Subunits, 29
Sympathetic ganglion, 43

T current, 11
Transmembrane signaling, 197
Turkey erythrocytes, 61

Xenopus oocytes, 77

Y1 cells, 133